作 者 简 介

类延菊，女，山东泰安人，博士，民盟盟员，湖南文理学院生命与环境科学学院教师。水产高效健康生产湖南省协同创新中心学术成员，湖南省双一流生物学专业成员。2014年获得中国海洋大学博士学位，研究方向为水产动物营养与饲料。毕业后来到湖南文理学院生命与环境科学学院，从事教学科研等工作。2009年至今，一直从事水产动物营养与重金属毒性关系的研究。博士期间在国家自然科学基金的资助下，开始研究重金属对水产贝类的毒性作用及抗氧化剂对重金属的解毒作用。工作之后继续研究此领域并获得湖南省自然科学基金1项、市级项目3项、校级项目1项，此方面研究结果中分别在 Aquaculture Nutrition、Aquaculture Research 等外文期刊中发表论文3篇，《动物营养学报》等中文期刊中发表论文3篇。

类延菊 ◎ 著

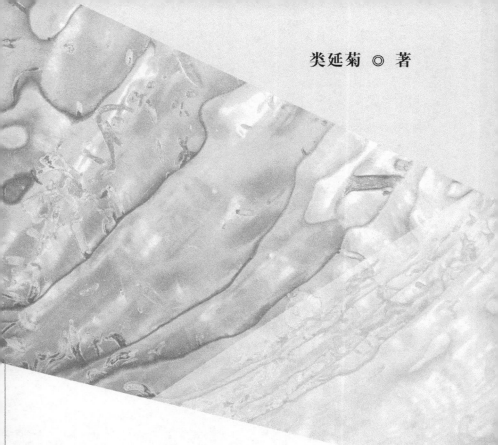

抗氧化剂对皱纹盘鲍
重金属的解毒作用研究

中国农业出版社
农村读物出版社
北 京

FOREWORD 前　言

重金属污染是近年渔业环境污染的公害之一，而且形势十分严峻。铜（Cu）和镉（Cd）是水域中最常见的两种重金属，其含量在我国的很多海域均超过了第四类国家海水水质标准。大量研究表明重金属对水产动物产生了严重危害，尤其是海水贝类很容易受到环境污染物的影响。过量的铜或镉都可以导致体内过量活性氧的产生，如不及时清除会危害细胞，引起脂质过氧化、糖和蛋白质硫醇氧化、DNA损伤等。需氧生物通过抗氧化系统来清除体内过量的活性氧，抗氧化系统和活性氧之间的动态平衡是保证机体健康状态的重要因素。如果平衡被打破，就会出现氧化应激，对生物体产生危害，尤其是对水生动物来说，这种平衡更容易被打破。所以减少氧化应激对生物体的危害，提高生物体自身的抗氧化能力，降低有害因素的刺激，一直是抗氧化研究的热点问题。

皱纹盘鲍（*Haliotis discus hannai* Ino）属于软体动物门腹足纲（Gastropoda）原始腹足目

(Archaeogastropoda) 鲍科 (Haliotidae)，是我国北方贝类养殖的主要品种之一，被誉为"海产八珍"之首。但是由于近年来环境污染的加剧，皱纹盘鲍不断暴发各类病害，产量大幅度下降，包括重金属在内的各种环境胁迫因子的增加导致病原微生物的高感染率和鲍的高死亡率。本研究选择海水中常见的重金属铜和镉，以皱纹盘鲍作为研究对象，探讨铜和镉对其抗氧化系统及氧化损伤的影响，为环境监测和海水贝类的健康养殖提供一定的理论依据。很多研究发现，饲料中添加抗氧化的物质可以提高生物体抗氧化的水平，从而降低氧化应激的胁迫。故根据"绿色、环保"的养殖策略，采用营养调控的方式来减轻重金属对皱纹盘鲍的毒性，并研究其解毒的调控机制。

本书内容包括：①选择水中的常见重金属铜和镉，研究水体中的铜和镉对皱纹盘鲍抗氧化指标及氧化损伤的影响，筛选皱纹盘鲍抗氧化敏感指标，一方面为环境监测提供可靠的数据，另一方面为皱纹盘鲍健康养殖提供参考。②在饲料中添加抗氧化剂硒和硫辛酸，研究能否降低铜或者镉对皱纹盘鲍的毒性，包括能否降低铜或镉在皱纹盘鲍组织的积累和减轻铜或镉造成的氧化损伤。在发现抗氧化剂一定程度上可以降低重金属毒性的基础上，进一步研究抗氧化剂解铜或镉毒性的机制，包括能否提高机体的抗氧化能力及金属硫蛋白的诱导表达。

前　言

本专著通过营养调控的方式来降低皱纹盘鲍重金属的毒性，并研究其解毒的调控机制，可为水产贝类重金属解毒提供一定的理论依据。但由于水平有限，疏漏之处在所难免，敬请广大读者批评指正。

著　者

2021年1月

CONTENTS 目 录

前言

第一章 重金属污染及抗氧化剂解毒的研究进展 /1

　　第一节　引言　　　　　　　　　　　　　　　　　　　　　　/1
　　第二节　铜和镉的污染现状　　　　　　　　　　　　　　　　/2
　　第三节　铜和镉的毒性作用　　　　　　　　　　　　　　　　/3
　　第四节　水产动物的抗氧化系统　　　　　　　　　　　　　　/6
　　第五节　重金属的解毒机制　　　　　　　　　　　　　　　　/10
　　参考文献　　　　　　　　　　　　　　　　　　　　　　　　/15

第二章 水中铜对皱纹盘鲍抗氧化反应、脂质过氧化和体内金属沉积的影响 /24

　　第一节　引言　　　　　　　　　　　　　　　　　　　　　　/24
　　第二节　摄食生长实验的设计　　　　　　　　　　　　　　　/25
　　第三节　铜浓度对皱纹盘鲍死亡率和抗氧化反应的影响　　　　/26
　　第四节　铜浓度对皱纹盘鲍脂质过氧化和体内金属沉积的影响　/32
　　参考文献　　　　　　　　　　　　　　　　　　　　　　　　/36

第三章 饲料硒解除铜对皱纹盘鲍毒性作用的研究 /40

　　第一节　引言　　　　　　　　　　　　　　　　　　　　　　/40
　　第二节　摄食生长实验设计　　　　　　　　　　　　　　　　/41

第三节 饲料硒对铜胁迫下皱纹盘鲍生长和组织铜
含量的影响 /43
第四节 饲料硒对铜胁迫下皱纹盘鲍肝胰脏抗氧化
指标的影响 /46
第五节 饲料硒对铜胁迫下皱纹盘鲍肝胰脏金属硫蛋白
（MT）含量和 MTF-1 mRNA 表达的影响 /50
参考文献 /60

第四章 饲料硫辛酸解除铜对皱纹盘鲍毒性作用的研究 /64

第一节 引言 /64
第二节 摄食生长实验设计 /65
第三节 饲料中的硫辛酸对铜胁迫下皱纹盘鲍生长
存活和组织铜含量的影响 /67
第四节 饲料中的硫辛酸对铜胁迫下皱纹盘鲍肝胰脏
抗氧化指标的影响 /70
参考文献 /75

第五章 水中镉对皱纹盘鲍抗氧化反应、脂质过氧化和
体内金属沉积的影响 /80

第一节 引言 /80
第二节 摄食生长实验设计 /81
第三节 镉浓度对皱纹盘鲍死亡率和抗氧化
反应的影响 /82
第四节 镉浓度对皱纹盘鲍脂质过氧化和
体内金属沉积的影响 /87
参考文献 /91

第六章 饲料硒解除镉对皱纹盘鲍毒性作用的研究 /96

第一节 引言 /96

目 录

　第二节　摄食生长实验设计　/98
　第三节　饲料硒对镉胁迫下皱纹盘鲍
　　　　　生长和组织镉含量的影响　/100
　第四节　饲料硒对镉胁迫下皱纹盘鲍肝胰脏
　　　　　抗氧化指标的影响　/102
　第五节　饲料硒对镉胁迫下皱纹盘鲍肝胰脏金属硫蛋白
　　　　　（MT）含量和 MTF-1 mRNA 表达的影响　/106
　参考文献　/108

第七章　饲料硫辛酸解除镉对皱纹盘鲍毒性作用的研究　/114

　第一节　引言　/114
　第二节　摄食生长实验设计　/115
　第三节　饲料中的硫辛酸对镉胁迫下皱纹盘鲍生长
　　　　　存活和组织镉含量的影响　/117
　第四节　饲料中的硫辛酸对镉胁迫下皱纹盘鲍肝胰脏
　　　　　抗氧化和氧化指标的影响　/119
　第五节　饲料中的硫辛酸对镉胁迫下肝胰脏金属硫
　　　　　蛋白（MT）含量、MT mRNA 和
　　　　　MTF-1 mRNA 表达的影响　/122
　参考文献　/125

第一章
重金属污染及抗氧化剂解毒的研究进展

第一节 引 言

随着全球经济的迅速发展,大量化工废水、生活垃圾、农药化肥被排放和使用,再加上自然侵蚀、风化,让很多的重金属流入水域。2008年,重金属从长江、珠江、黄河等河流入海的总量超过3 000t。重金属污染具有易被生物富集、生物放大效应和危害性大等一系列特点。水中重金属的增加不仅污染水环境,而且严重危害人类及其他生物的生存。

重金属的污染物一般分为两类,一类为生物体所必需的微量元素,如铜、锌、钴、锡等,这些元素一旦过量,可成为水中污染元素,对水生生物产生毒害作用;另一类并不具备生理功能,但是这些金属可以显著影响生物体生长存活,如镉、铅、汞、铬等。重金属可以集聚在藻类和污泥中,接着被鱼类、贝类和虾蟹类等体表吸附或吸收入体内,因而造成污染。重金属离子会对水生生物细胞膜造成损害,抑制细胞分裂,导致生殖异常甚至死亡(Brya, 1984)。其中,铜(Cu)和镉(Cd)是不同类型水中常见的重金属,并且在水域中的污染越来越严重。

第二节 铜和镉的污染现状

铜（Cu）是一种体内必需的微量元素，对生物体生长和发育起着重要的作用。低浓度的 Cu 对于生物体的作用非常重要，是生物体的营养元素，而且 Cu 是水生无脊椎动物血液氧的载体——血蓝蛋白的中心原子，所以它在很多生物体生物酶的结构和功能上发挥着非常重要的作用。但过量的 Cu 对水生生物是有害的（McGeer 等，2000；Beyers 和 Farmer，2001；Shariff 等，2001），甚至是可致死的（程波等，2008）。根据 Yeh 等（2004）的研究表明，高浓度的 Cu 会影响水生动物的血细胞数量及一些氧化酶的活性。另外，Bambang 等（1995）还发现 Cu 的氧化胁迫可以影响水生生物渗透压的调节及生存。过量的 Cu 对人体有害，中国农业行业标准《无公害食品 水产品中有毒有害物质限量》（NY 5073—2006）规定，所有水产品 Cu 的含量应小于 50mg/kg。近年来，由于经济的快速发展，大量工农业废水废弃物直接排放入海中，导致海水中的重金属和沉积物中的污染问题日益突出。尤其是 Cu 的污染已经越来越严重，比如在水产养殖生产管理过程中通常会采用 0.5mg/L 的 $CuSO_4 \cdot 5H_2O$ 溶液来杀藻杀菌，以及投入人工配合饵料和水质改良剂及肥料的过程均会带来 Cu 的污染（Yeh 等，2004；程波等，2008）。根据 2006 年的《中国渔业生态环境状况公报》报告，我国天然重要渔业水域中 12.8% 的监测面积受到了不同程度的 Cu 污染。

在美国毒物管理委员会列出的危害人体健康的有毒物质中，镉（Cd）排在第 6 位。随着工农业经济的发展，受污染水域中的 Cd 含量逐年上升（刘杰，1998）。黄浦江表层沉积物中 Cd 超过背景值 2 倍，苏州河 75% 的采样点 Cd 超标。沿海海域中，大连湾 Cd 含量在 60% 的监测站中超标，锦州湾河口沉积物 Cd 的含量已经超过了三类海洋沉积物标准[《海洋沉积物质量》（GB 18668—2002）]。Cd 是生物体生长的非基本元素，具有非常大的生物毒性，Cd 半衰期长达 10~20 年，为目前已知的最容易在生物体内蓄积的元素（刘杰，1998）。Cd 是一个很重要的环境污染物，具有"三致"效应，即致癌、致畸、致突变，威胁着人类的健康（Zenzes 等，1995）。Cd 可以积累在生物体的许多组织中（Friberg Led，1979），可以在人体的肾脏和肝脏中积累并引起中毒，可能会导致癌症，如 Cd 胁迫可以诱导睾丸间质细胞癌（Waolkes 等，1985）。日本的富山县发生在 20 世纪 60 年代的"骨痛病"，具体原因就是当地居民食用了被 Cd 污染的稻米，该病因引发周身剧痛而得名（杨世勇等，2000）。

第三节 铜和镉的毒性作用

研究表明，重金属主要通过诱导细胞的脂质过氧化引起细胞毒性的产生。在细胞氧化胁迫的过程中产生过量的自由基，这些自由基可以攻击细胞的蛋白质、脂质和 DNA 等（Yousef 等，2007；Kasperczy 等，2008；Guzmán 等，2009）。此外，还有实验研究表明，Cu 通过被动运输或主动运输的方式被吸收进入生物体内，与生物体内的多种酶类或活性因子如硫基、胺基、亚胺基等结合而形成不溶水的硫醇盐，使酶或活性因子失活，破坏细胞的酶系统，进一步阻断细胞的正常的生化反应和新陈代谢，甚至会导致生物个体的死亡（谢炎福和祖恩普，2005）。

现在有很多理论来解释 Cu 诱导细胞毒性的机制。大多数情况下，这些理论机制都认为铜离子参与活性氧自由基（ROS）的形成。Cu^{2+} 和 Cu^+ 在生物体内可以参加氧化还原反应。在过氧化物、还原剂如维生素 C 或谷胱甘肽（GSH）存在时，Cu^{2+} 被还原成 Cu^+，在这个过程中通过 Haber/Weiss 反应将 H_2O_2 催化生成 ·OH（Bremner，1998；Kadiiska 等，1993），具体的反应式为：$2O^{2-} + 4Cu^{2+} \rightarrow O_2 + 4Cu^+$，$Cu^+ + H_2O_2 \rightarrow Cu^{2+} + OH^- + ·OH$。羟自由基（·OH）是生物生命系统中所有的自由基中危害性最强的自由基。它几乎能与所有的生物分子发生反应（Buettner，

1993），可以诱导氧化蛋白质 C 端或不饱和脂肪酸的氢，从而产生碳为中心的蛋白质自由基和脂质自由基（Powell，2000）。Cu 通常与大分子的物质（如蛋白质、DNA、多糖）关联，适量的 Cu 可以表现出较好的抗氧化性（Lauridsen 等，2000）。然而，若 Cu 积累超过一定的限度，它可以通过氧化还原反应产生活性自由基（Powell，2000）。因为 Cu 主要在肝脏中积累，ROS 主要在实验动物的肝脏中产生（Kadiiska 等，1993）。

过量 Cu 最明显的毒性就是可以诱导细胞膜的脂质过氧化作用，在脂质过氧化反应中会产生脂质自由基和氧，从而形成过氧自由基（Powell，2000）。脂质过氧自由基可以通过改变细胞膜的流动性和渗透性，或直接攻击 DNA 和其他细胞内的大分子，如蛋白质（Mattie 和 Freedman，2001），对细胞造成损伤。脂质过氧化反应发生在线粒体中（Britton，1996），Cu 还可以导致肝细胞溶酶体膜的过氧化作用（Sokol 等，1990；Bremner，1998）。过量 Cu 处理下的大鼠出现氧化损伤，包括肝脏 α-生育酚和 GSH 水平降低，线粒体脂质过氧化产物增加，肝线粒体呼吸的控制率减少，细胞色素 c 氧化酶的活性降低（Sokol 等，1990）。研究表明，随着大鼠肝脏 Cu 含量的增加，脂质过氧化的产物丙二醛（MDA）的含量显著升高（Ohhira 等，1995；Sansinanea 等，1998）。在 Cu 的胁迫下，小鼠的肝脏和血清中的谷丙转氨酶和 MDA 的含量显著性升高，红细胞和全血铜锌超氧化物歧化酶（CuZn-SOD）和硒谷胱甘肽酶（Se-GPx）的活性显著性降低。Cu 的超负荷降低了细胞色素 c 氧化酶的活性，损害肝脏线粒体呼吸（Myers 等，1993），从而导致肝脏毒性。硫酸铜注射后大鼠的过氧化氢酶（CAT）和谷胱甘肽过氧化物酶（GPx）的活性显著性降低（Ossola 等，1997）。

Cd 也具有非常强的诱导细胞膜脂质过氧化反应的作用（刘瑞明等，1990）。细胞膜脂质过氧化的过程中可以产生很多活性氧自由基，主要包括超氧阴离子（O^{2-}）、羟基自由基（·OH）、过氧化物自由基（ROO·）、烷氧自由基（RO·），这些活性氧自由基可以使DNA发生氧化损伤、氧化蛋白质和消耗ATP储存等,并可以诱导细胞凋亡。研究表明,Cd可以抑制超氧化物歧化酶(SOD)的活性,导致活性氧水平升高、诱导膜脂质过氧化及损伤神经细胞(Shukla等,1989)。因为Cd属于非典型的过渡金属元素,它不易因为电子的转移而诱发脂质过氧化(赵燕等,1995)。研究表明,Cd对细胞内的巯基及抗氧化酶的破坏是导致脂质过氧化的原因(Sharma等,1991)。机体的抗氧化系统被破坏后会使细胞自身清除自由基的能力降低,机体自由基浓度将会增高。用含Cd 250mg/L的水处理大鼠4周以后,发现大鼠的血清和肝组织中MDA的含量显著增高,GPx的活性显著降低,生物体的抗氧化功能下降,机体的脂质过氧化明显。

研究还表明,Cd还可以引起DNA的氧化损伤,使DNA的单链断裂,氧化生成碱基的修饰产物8-羟基脱氧鸟苷(Koizumi等,1992;Mikhailova等,1997)。Ochi等(1983)研究证实,Cd引起的DNA断裂过程需要氧的参与,如果在反应过程中加入抗氧化酶SOD,单链DNA的断裂概率将会显著下降,可以得出氧自由基在Cd引起的DNA断裂的过程中发挥了非常重要的作用。研究表明,Cd不仅降低了蛋白质和RNA的合成速度,并且破坏了细胞的核糖核蛋白体(王渊源,1993)。

总之,重金属一方面通过多种途径可以在生物体内诱发大量的自由基代谢物和活性氧产生;另一方面直接作用于细胞的抗氧化酶系统,降低细胞清除氧自由基的能力。在这个过程中可以使机体内产生大量活性氧自由基,使细胞的DNA断裂、脂质过氧化、酶蛋白失活等,从而引起机体氧化应激,影响机体的代谢活动。

重金属对水产动物产生毒性作用的原因也是诱导有机体产生过量ROS,从而导致生物体的氧化应激,引起氧化损伤。抗氧化体系是消除ROS产生的主要机制。研究表明,水生动物抗氧化系统与重金属的浓度之间存在相关性(Lopes,2001)。在重金属氧化胁迫作用下,牙鲆的鳃组织中SOD、CAT和GPx的活力表现出低浓度诱导、高浓度抑制的规律（王凡等，2006）。根据王凡等（2008）研究表明，在 Cu 处理下，扇贝的内脏团中的抗氧化酶（SOD、CAT、GPx）的活力表现为先抑制后诱导然后又抑制的变化规律，但是从总体上来说表现为抑制。不同浓度的 Cu 处理黄河鲤，低浓度的 Cu 对总抗氧化能力（T-AOC）起激活的作用，但是高浓度的 Cu 却抑制 T-AOC，究其原因可能和抗氧化酶的活性受到抑制有关（高春生等，2008）。用不同浓度的 Cu 处理黑鲷时，随着水体中的 Cu 浓度的逐渐升高，抗氧化酶 SOD 和 CAT 活性逐渐被抑制，并且其活性随着 Cu 浓度的升高而降低（沈盎绿等，2007）。总之，重金属对水生动物的抗氧化系统表现出一个低浓度诱导、高浓度抑制的变化规律，这与其他重金属离子的影响规律具有相似性（Jing 等，2006）。

第四节　水产动物的抗氧化系统

重金属可以诱发水生生物产生过量的活性氧，为了对抗这种氧化胁迫，水生生物在长期自然进化的过程中形成了完整的抗氧化系统，这套抗氧化系统是消除机体过量的活性氧的主要机制。ROS 的种类主要有超氧阴离子（O^{2-}）、羟基自由基（·OH）、单线态氧（$^1O_2$）、过氧化氢（H_2O_2）、氧化物自由基（ROO·），以及处于激发态的氧等（Bartosz，2009）。抗氧化系统和 ROS 之间的动态平衡是保证生物体健康的重要因素。若生物体内的 ROS 水平高于其自身的抗氧化系统的清除能力，那么过多的 ROS 将会通过化学链的反应破坏体内的一些物质（如脂质过氧化、蛋白质和 DNA 损伤），进一步损伤细胞结构和功能，甚至导致细胞的死亡（图 1-1）（Nordberg 等，2001；Fang 等，2002；Murphy 等，2006）。

生物体在消除 ROS 的过程中主要通过以下几道防线：首先，从根本上避免生物体内 ROS 的产生，主要是依靠一些氧化酶（如细胞色素氧化酶等）完成。细胞色素氧化酶可以显著减轻超氧阴离子和过氧化氢对细胞的危害（Fridovich，1989）。其次，机体过量的

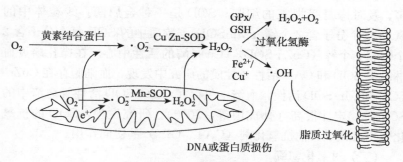

图 1-1 自由基对机体的损伤

ROS 产生之后，机体的一些抗氧化酶和抗氧化的小分子来清除过量的自由基。例如，超氧化物歧化酶（SOD）可以催化超氧阴离子生成 H_2O_2 和氧分子；过氧化氢酶（CAT）和谷胱甘肽过氧化物酶（GPx）均可以催化 H_2O_2 形成水和氧分子。再次，通过抗氧化剂来阻止呼吸链的激发和延伸（Fridovich，1989）。最后一道防线是 ROS 对生物体造成一定的损伤之后损伤的修复，如被氧化的蛋白质在相关蛋白酶的作用下降解、过氧化脂质的修复等。

目前抗氧化系统被分为三大类：①体内可以清除或转化 ROS 的抗氧化酶系统，主要包括 SOD、CAT、GPx、谷胱甘肽硫转移酶（GST）、谷胱甘肽还原酶（GR）、硫氧还蛋白还原酶（TrxR）、硫氧还蛋白过氧化物酶（TrxP）（Bogdan 等，2000；Nordberg 等，2001；Abele 等，2004）。②非酶的抗氧化小分子物质，这类小分子物质在细胞内合成，不需要基因的编码。如谷胱甘肽（GSH）、硒、α-硫辛酸（LA）、维生素 E、维生素 C 等。③动物体内一些功能蛋白，其具有结合自由基的功能，如金属硫蛋白（MT）、硫氧还蛋白（Trx）、硒结合蛋白（SeBP）等（Song 等，2006；Bigot 等，2009）。最新研究发现了一些小分子抗氧化多肽（Yang 等，2009）。

（一）超氧化物歧化酶

超氧化物歧化酶（SOD）广泛存在于动植物和微生物中，是目前发现的唯一以超氧阴离子自由基为底物的酶，是生物体内最重要的抗氧化酶。它可以阻止和对抗活性氧自由基对细胞造成的危害，及时修复受损害的细胞。SOD 是一种金属酶。线粒体中的 SOD 每个分子含 2 个锰（Mn-SOD），胞质中的 SOD 每个分子含 2 个铜和 2 个锌（CuZn-SOD），Cu 是酶的活性中心。在栉孔扇贝血淋巴中 SOD 和 CAT 活性及性质的研究中发现，血细胞存在 CuZn-SOD 和 Mn-SOD 活性，血清中仅存在 CuZn-SOD 活性且血清中的 SOD 热稳定性很高（孙虎山等，2000）。研究发现，SOD 的抗氧化作用有赖于其他抗氧化剂（CAT、GPx）的协同作用。

（二）过氧化氢酶

过氧化氢酶（CAT）也称为触酶，是体内另一种非常重要的抗氧化酶，最主要的作用就是参与分解清除细胞器中所产生的 H_2O_2，使过氧化氢分解为水和氧气。Stern（1936）证明，卟啉环是 CAT 活性中心。在哺乳动物中，肝脏的 CAT 含量最高，结缔组织中含量最低。Mourente 等（2002）认为从 CAT 和 SOD 的作用机制可以推测这两种酶的活力变化应该保持同步性。CAT/SOD 是动物体内的一个重要的抗氧化指标，它的升高标志着抵制 ROS 的抗氧化酶的激活，也表示动物体受到的氧化压力在增大。研究还发现，无脊椎动物的 SOD 和 CAT 活性比脊椎动物要高，目前还不确定这种现象的具体意义，但可以说明 SOD 和 CAT 在水产无脊椎动物中具有重要作用。

（三）谷胱甘肽系统

谷胱甘肽过氧化物酶（GPx）、谷胱甘肽还原酶（GR）、谷胱甘肽硫转移酶（GST）是还原性谷胱甘肽（GSH）依赖型的抗氧化酶，因为它们都以 GSH 为辅酶，所以称这个系统为谷胱甘肽抗氧化系统（Meister 和 Anderson，1983）。前两种是抗氧化防御系统的酶，GST 属于 Ⅱ-相转移酶。GPx 对 GSH 具有较高的亲和性，而且可以将 GSH 氧化为 GSSH。GPx 对脂肪氧化过程中产生的过氧化氢和有机物的过氧化物具有解毒的作用（Verlecar 等，2008）。GPx 又包括依赖硒的和不依赖硒的两种，依赖硒的 GPs 过氧化物酶可以和一系列的包括过氧化氢和有机过氧化物在内的过氧化物起反应，不依赖硒的 GPx 不能以过氧化氢为底物起反应。不依赖硒的 GPx 也被证明过和 GST 酶家族中的一个酶相同（Ceballos-Picot 等，1996）。GPx 和 CAT 都具有清除过氧化氢的作用，但是在研究外来大型红藻对双壳类氧化应激的实验中发现，SOD 产生的过氧化氢是由 GPx 来清除的而不是 CAT（Box，2009）。分析这种原因主要是 GPx 对过氧化氢的亲和力较高，所以在生物体内对过氧化氢主要通过 GPx 来清除，但是 CAT 的米氏常数（K_m）的值较大，所以 CAT 比 GPx 能更有效地清除体内高浓度过氧化氢。GR 是一种黄素蛋白氧化还原酶，能催化 GSSH 还原成 GSH，而 GSH 是 GPx 发挥作用的必要物质且 GSH 具有重要的抗氧化作用。Ⅱ-相转移酶 GST 可以使某些内源或外源性的亲电子基团与 GSH 的巯基偶联，从而增加其疏水性，使其容易穿越细胞膜，分解排出体外，以达到解毒的作用。软体动物中谷胱甘肽抗氧化系统一直是研究的热点，主要在其抗氧化作用和抗重金属等的解毒作用及作为环境污染的生物标志物方面。

（四）硫氧还蛋白系统

硫氧还蛋白系统包括硫氧还蛋白还原酶（TrxR）、硫氧还蛋白（Trx）和还原型烟酰胺腺嘌呤二核苷磷酸（NADPH），在生物体氧化还原的调节及抗氧化防御过程中起非常重要的作用。其中，TrxR 是一种含硒酶，它的活性与生物体内的硒含量紧密相关，主要功能为可以再生小分子蛋白质 Trx（Takashi，1996）。TrxR/Trx 系统是生物体内的非常重要的巯基/二硫键还原系统。Trx 是分子质量为 12ku 的小分子蛋白质，在组织对抗氧化反应的过程中起着重要的作用（Powis 等，2001）。

（五）抗氧化小分子

抗氧化系统除了抗氧化酶之外还包括一些水溶性还原剂（GSH、抗坏血酸、尿酸盐）、脂溶性的还原剂（生育酚、β-胡萝卜素）。GSH 和抗坏血酸在某些程度上是矛盾的，因为它们既可以作强氧化剂，也具有抗氧化的作用，这些水溶性的还原剂在细胞膜中不能发挥自身抗氧化的功能。生育酚不仅可以清除可能引发脂质过氧化连锁反应中的氧自由基，而且可以清除脂质氧化链中的自由基。这也就意味着抗坏血酸和生育酚对细胞膜的抗氧化的保护具有协同作用。小分子维生素 E、维生素 C、维生素 A、GSH、类胡萝卜素、血浆铜蓝蛋白、尿酸、葡萄糖、Se、Zn、Cu、Fe 等组成抗氧化的第一道防线（Klaunig 和 Kamendulis，2004；Limón-Pacheco 和 Gonsebatt，2009）。

第五节 重金属的解毒机制

无论是 Cu 还是 Cd 都能够催化机体活性氧的产生，所以具有抗氧化能力的营养物质通过提高机体的抗氧化能力来保护重金属导致的氧化损伤。这些抗氧化剂的抗氧化机制主要包括抑制线粒体超氧化物的产生、活性氧自由基的清除、从产生活性氧的部位螯合出金属离子、减少氢的过氧化物的生成、修复受损的分子（Chow，1991）。现在有许多不同类型的抗氧化添加剂，由于有些抗氧化剂存在毒副作用，如残留、抗药性等问题，所以研究和开发安全、高效、无毒副作用的绿色添加剂具有重要意义。常用的天然抗氧化添加剂主要包括：①微量元素，如硒、铁、铜、锌和锰等。②维生素，如维生素 A、维生素 E、维生素 C、维生素 D 和 β-胡萝卜素等。③非营养性添加剂，如硫辛酸、GSH、虾青素、茶多酚，还包括一些中草药添加剂等。研究证明，添加抗氧化剂在一定程度上可以提高水产动物的抗氧化水平。以下就抗氧化剂硒和硫辛酸解重金属毒性的研究进展进行综述。

（一）硒的解重金属毒性的机制

硒（selenium，Se）是广泛存在于动植物体的非金属元素，而且是动植物体所必需的微量元素。硒是一种天然的抗氧化剂，是抗氧化酶 GPx 和 TrxR 的活性中心。在生物体内，硒主要以硒半胱氨酸的形式与重要的蛋白质结合，来行使其各种功能。在水产动物中，目前研究较多的是其在正常环境条件下的免疫和抗氧化作用，在氧化胁迫下的硒的抗氧化作用研究较少。硒拮抗重金属的毒性作用包括3个方面（王迎红，2000）：①硒增强了细胞抗氧化酶的活性，对重金属氧化胁迫产生的活性氧自由基具有较强的清除能力，从而防止自由基对生物大分子的攻击。②硒可以诱导解毒蛋白——金属硫蛋白（MT）的合成，金属硫蛋白和重金属螯合而降低重金属对机体的危害。③硒和重金属结合成硒-重金属复合物，减轻重金属的危害。

1. 硒的抗氧化作用　硒是体内抗氧化体系的组成部分，常作为含硒抗氧化酶（如 GPx 和 TrxR 等）的活性中心而参与抗氧化作用（Rotruer 和 Poue，1993；Maleki 等，2005）。Newairy 等（2007）和 Tran 等（2007）研究发现，硒的添加增强了机体抗氧化作用，从而降低了 Cu 或 Cd 的毒性。许多研究发现，硒可以通过调节 GSH 的含量和抗氧化酶的活力而降低脂质过氧化的毒性（Combs 和 Combs，1984；McPherson，1994；Ji 等，2006）。研究表明，硒可以提高氧化损伤的肝细胞 SOD 和 GPx 活性及 GPx mRNA 表达水平，还可以降低脂质过氧化产物 MDA 的含量，在一定程度上可以保护氧化损伤的肝细胞。Trevisan 等（2011）研究表明，在 Cu 处理之前在水中预加一定量的硒，从很大程度上可以抵御紫贻贝（*Mytilus edulis*）蛋白疏基的过氧化和 DNA 的损伤。杨成峰等（1996）研究发现，硒+镉组合组的大鼠肝微粒体、线粒体内脂质过氧化物水平显著降低，并且 GPx 活性高于对照组和单纯染镉组。廖琳（2002）报道，硒的添加增加了 GSH 的形成。GSH 可以将亚硒酸盐还原为硒化合物，硒化合物具有高度的亲脂性，因而改变了 Cd 在关键组织中的分布及毒性。

2. 硒通过诱导金属硫蛋白的合成解重金属毒性　硒解重金属毒性的另一条途径为硒能诱导金属硫蛋白（MT）的合成。郭军华等（1993）研究表明，硒也能诱导 MT，且目前公认其比对 MT 诱导能力较强的 Cd 的作用还好。后来用酶联免疫吸附法也证实了硒对 MT 的诱导作用，而且诱导效果为 Zn 的 8～9 倍。李盟军等（1995）证明了亚硒酸钠可以诱导金属硫蛋白，抵抗顺铂所致小鼠的生殖毒性。动物注射 Cd 或 Zn 盐能导致 Cd-MT 和 Zn-MT 在组织中的积累，这是由于 MT 的基因转录受离子诱导激活后，可以转录出大量的 MT-mRNA，从而指导合成多量 MT（周杰昊等，1995）。所以 MT 对过量的 Zn 和 Cu 有解毒作用，当 Zn 和 Cu 水平过量时，离子就以 MT 结合态的形式存在，这样有利于从组织中将这些金属清除出来，起到了解 Zn 和 Cu 毒害的作用。王宗元等（1997）研究证实，硒可明显改变氯化镉在组织中的分布，使已和金属硫蛋白结合的镉剥离，从而有利于镉的排出。

3. 金属硫蛋白解重金属毒性　金属硫蛋白（MT）是一类广泛存在于生物体内的富含半胱氨酸、可以和金属结合的低分子质量蛋白。MT 生物学活性很广泛，主要包括：①参与多种微量元素的代谢，如 Zn、Cu 等；②解除或降低 Cu、Cd、Hg、Pb 等重金属的毒性；③对抗自由基，减少生物氧化及抗衰老；④减轻生物的应激反应，增强对各种不良状态的适应；⑤抗肿瘤。MT 可以与 Cd、Cu 等有毒的重金属离子结合以减少其对组织的直接氧化损伤，可以调节机体抗氧化能力及解重金属毒性（加春生等，2007）。MT 是生物体内的重要应激保护蛋白。大量的研究证明，MT 优于其他抗氧化剂的抗氧化作用（赵红光和龚守良，2005）。Se 的制剂是生物体合成 MT 的有效诱导物（田晓丽等，2006；郭家彬等，2006），据推测，Se 诱导的 MT 合成的增加，对 Cd 或 Cu 毒性有一定的拮抗作用。此外，MT 含有大量的低能的空轨道和巯基，可以清除多种氧的自由基，并且 MT 中的巯基基团可以使受损的 DNA 修复。硒化合物可以封闭 MT 与 Cd 的结合部位，从而激活分子质量更大的蛋白质，夺取细胞液中 Cd，使和 MT 结合的 Cd 剥离，从而有利于 Cd 被排出体外（王宗元等，2007）。但是，有一些学者认为，虽然细胞内 Cd-MT 对生物体具有保护作用，但在生物体的细胞外，Cd-MT 比 Cd 本身具有更大的肾细胞毒性，因此 MT 对重金属是否有解毒作用还存在争议。

此外，金属硫蛋白 MT 可以消除重金属氧化应激过程中产生的自由基，在抑制脂质过氧化的过程中发挥重要的作用。Thornalley 和 Vasak（1985）研究表明，MT 具有很强的消除自由基能力，可以保护细胞膜的突变。李兆萍等（1989）的研究证明，MT 可以显著减轻细胞缺氧/复氧的损伤，提高大鼠心肌细胞存活率，显著降低细胞中钙的积累，抑制细胞膜脂质过氧化。MT 还是一种非常有效的自由基捕获剂。MT 不仅能清除 O_2^-，而且对·OH 也有很强的清除能力，且捕获·OH 的能力比 O_2^- 要强，MT 捕获·OH 的能力是 GSH 的 100 倍（帖建科等，1995；张亨山，1994）。MT 对 O_2^- 的捕获能力不如 SOD，但是 SOD 不能清除·OH，说明 MT 在清除自由基方面比 SOD 要强。MT 在清除自由基的过程中，可以释放微量元素 Zn，提高自我修复机能（陆超华，1995；奇云等，1995）。因此，MT 对·OH引起的 DNA 损伤有一定的保护作用。

4. 金属硫蛋白转录调控机制 关于贝类体内对重金属的调控机制，研究最多的是金属调控转录因子（metal regulation of transcription，MTF-1）。MTF-1 是目前为止发现的唯一的重金属转录调控因子。MTF-1 中的锌指结构可以和金属离子（Zn，Cd）结合，引起其构象的转变，转入细胞核中，与效应基因的上游 MRE（metal response element）结合，对效应基因进行调控，从而调控基因的表达（Langmade 等，2000）。目前对 MTF-1 基因的研究主要在人、小鼠、果蝇中，在无脊椎动物的研究还很少，全长序列也鲜有报道。已有研究表明，Zn、Cu、Cd 等金属离子能够诱导 MTF-1 基因的转录调控，而且诱导机制相似。但是与 MTF-1 结合的主要是 Zn，而 Cd 和 Cu 是通过取代 Zn 在金属硫蛋白上的结合点，使 Zn 游离出来，调控 MTF-1 的转录，从而调控下游基因（Andrews 等，2001）。在贝类中 MTF-1 转录调控的靶基因筛选还未开展研究。

（二）硫辛酸解重金属毒性的机制

1. 清除重金属胁迫产生的 ROS 硫辛酸（lipoic acid，LA）是双硫五元环的维生素，化学名称为 1,2-二硫戊环-3-戊酸（图 1-2），在生物体内可以转化为还原型的二氢硫辛酸（DHLA），被誉为"万能的抗氧化剂"（Packer 等，1995）。研究表明，LA 可由八碳的脂肪酸和硫元素在线粒体中合成（Packer 等，2001）。

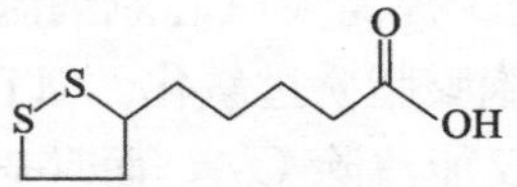

（相对分子质量：206.33）

图 1-2　α-LA 的结构

　　LA 可以直接清除体内氧化胁迫过程中产生的 ROS，如 ·OH、H_2O_2、单线态氧（$^1O_2$）、过氧化亚硝基（·ONOO）和次氯酸（HClO）等（Gerreke 等，1997），还可以阻断氧化还原反应链。LA 和 DHLA 在生物体内氧化还原的过程中可以再生一些其他的抗氧化剂，例如 GSH、维生素 E、维生素 C、Trx 及泛醌等，可以将这些抗氧化剂由氧化型转化为还原型。很多的研究发现，LA 具有强大的抗氧化能力，在预防和治疗一些氧化胁迫相关的疾病中发挥积极的作用（Biewenga 等，1997；Çakatay，2006）。

　　2. 增强体内抗氧化酶的活性　　研究表明，LA 可以增加体内抗氧化酶的活性来发挥其抗氧化的功能。动物体内抗氧化酶在抵抗氧化胁迫的过程中发挥着重要的作用，它们可以催化不同活性氧的反应，可以使活性氧变成 H_2O 和 CO_2。Arivazhagan 等（2001）研究表明，在大鼠日粮中添加 100mg/kg LA 饲喂 7～14d 后，可显著升高组织中抗氧化酶 SOD、GPx、GR 的活性。Leonard 等（1999）报道，在重金属造成小鼠耳蜗氧化损害后，饲喂 LA 后，抗氧化酶 CAT、SOD、GPx、GR 的活性显著提高，且表现出剂量依赖效应。根据 Carey 等（2002）研究表明，在纯种马的日粮中添加 10mg/kg 的 LA，可以显著提高白细胞中 GPx 水平。Kowluru 等（2005）研究证明，糖尿病导致的视网膜功能异常的小鼠的线粒体 SOD 活性受到显著抑制，但是在给小鼠补充 400mg/kg LA 11 个月后，线粒体中的 SOD 活性上升近 30%，并且其 mRNA 水平上升了 25% 左右。Park 等（2006）研究表明，在饲料中添加 1 000mg/kg 的 LA 后，鲷的肌肉组织具有再生维生素 C 的功能。根据陈齐勇等

（2011）研究表明，添加 LA 可以显著提高皱纹盘鲍的抗氧化水平。目前来看，对于 LA 和 DHLA 提高这些抗氧化酶活性的机制还不清楚，可能是 DHLA 的巯基结构提高了巯基酶活性的缘故。在研究 LA 抗 Cd 的实验中发现，LA 可完全消除 Cd 诱导的大脑脂质过氧化，恢复碱性磷酸酶和 CAT 的活力，而且使脑的 GSH 恢复到正常水平（Lester，1995）。

3. 螯合金属离子 生物体内 Cu、Cd 等重金属能够激活体内过量 ROS 的产生，从而对机体的大分子物质产生损伤，损害机体的机能。研究表明，LA 可以通过某种方式螯合或吸附金属离子，阻止重金属离子产生自由基，降低其过氧化的程度。

参 考 文 献

陈齐勇，张文兵，麦康森，等，2011. 饲料中硒和 α-硫辛酸对皱纹盘鲍稚鲍抗氧化反应的影响 [J]. 中国海洋大学学报：自然科学版，41（5）：38-42.

程波，刘鹰，杨红生，2008. Cu^{2+} 在凡纳滨对虾组织中的积累及其对蜕皮率、死亡率的影响 [J]. 农业环境科学学报，27（5）：2091-2095.

高春生，王春秀，张书松，2008. 水体铜对黄河鲤肝胰脏抗氧化酶活性和总抗氧化能力的影响 [J]. 农业环境科学学报，27（3）：1157-1162.

郭家彬，陈立娟，闫长会，等，2006. 硫酸锌诱导金属硫蛋白对阿霉素致心脏氧自由基产生的影响 [J]. 中国临床药理学与治疗学，11（7）：725-729.

加春生，李金龙，徐世文，2007. 镉致鸡血管氧化应激与金属硫蛋白含量的变化 [J]. 生态毒理学报，2（2）：178-183.

李盟军，郭军华，徐卓立，等，1995. 甘草锌、富硒麦芽和亚硒酸钠诱导金属硫蛋白对顺铂所致生殖毒性的影响 [J]. 军事医学科学院院刊（3）：239-240.

李兆萍，1989. 金属硫蛋白对大鼠心肌细胞缺氧-复氧损伤的保护作用 [J]. 科学通报，34（7）：544-546.

廖琳，胡晓荣，李晖，等，2002. 生态环境中镉对生物体毒性作用机理及硒对该毒性拮抗作用的研究进展 [J]. 四川环境（2）：21-24.

刘杰，1998. 镉的毒性和毒理学进展 [J]. 中华劳动卫生职业病杂志，16（1）：2-4.

刘瑞明，1990. 镉的肝细胞毒性与脂质过氧化关系的研究 [J]. 中国环境科学，

10 (3): 187-190.

陆超华, 1995. 金属硫蛋白的生理功能及应用 [J]. 广州食品工业科技, 11 (4): 21-24.

奇云, 祁向东, 1995. 令医学界刮目相看的金属硫蛋白 [J]. 知识介绍 (10): 27-29.

沈盎绿, 马胜伟, 贾秋红, 2007. 黑鲷受铜、镉胁迫的生理反应 [J]. 海洋渔业, 29 (3): 257-262.

孙虎山, 李光友, 2000. 栉孔扇贝血淋巴中超氧化物歧化酶和过氧化氢酶活性及其性质的研究 [J]. 海洋与湖沼, 31 (3): 259-265.

唐朝枢, 李兆萍, 苏静怡, 1989. 金属硫蛋白是机体内源性抗损伤物质 [J]. 北京医科大学学报, 21 (2): 108.

田晓丽, 唐欣, 左刚, 等, 2006. 硒与金属硫蛋白对小鼠肝损伤的防护作用研究 [J]. 中国生物工程杂志, 26 (6): 23-29.

帖建科, 李令媛, 茹炳根, 等, 1995. 金属硫蛋白清除自由基及其对自由引起的核酸损伤保护作用的研究 [J]. 生物物理学报, 11 (2): 276-282.

王凡, 李法松, 赵元凤, 等, 2006. 铜对扇贝肌肉抗氧化防御系统的影响 [J]. 安徽农业科学, 34 (23): 6109-6110, 6112.

王凡, 赵元凤, 吕景才, 等, 2008. 铜污染对扇贝内脏团抗氧化酶活性的影响 [J]. 水产科学, 27 (12): 622-624.

王迎红, 胡国刚, 刘绣, 2000. 硒拮抗砷抑制体外人胚肺组织中抗氧化酶活性的研究 [J]. 中国地方病学杂志, 19 (2): 87-89.

王渊源, 1993. 鱼虾营养概论 [M]. 厦门: 厦门大学出版社.

王宗元, 史德浩, 卞建春, 等, 1997. 亚硒酸钠防治镉中毒的分子机理 [J]. 中国药理学与毒理学杂志, 11 (2): 114-115.

谢炎福, 祖恩普, 2005. 硫酸铜引起鱼类中毒原因的分析及对策 [J]. 水利渔业, 25 (6): 98-99.

杨成峰, 陈学敏, 1996. 硒对镉在大鼠肝脏亚细胞分布及脂质过氧化作用的影响 [J]. 卫生研究, 25 (1): 32-34.

张亨山, 1994. 金属硫蛋白的抗氧化作用 [J]. 国外医学卫生分册, 21 (2): 21-25.

赵红光, 龚守良, 2005. 金属硫蛋白抗氧化损伤作用及其机制 [J]. 吉林大学学报: 医学版, 31 (2): 322-324.

赵燕, 江文, 侯京武, 等, 1995. 钙负荷和丹参酮对线粒体脂质过氧化的影

响 [J]. 生物化学与生物物理学报, 27 (6): 610-615.

ABELE D, PUNTARULO S, 2004. Formation of reactive species and induction of antioxidant defense systems in polar and temperate marine invertebrates and fish [J]. Comp. Biochem. Physiol. A, 138: 405-415.

AL-AKEL-AS, SHAMSI MJK, AL-KAHEM HF, et al, 1988. Effect of cadmium on the cichildfish, *Oreoehromis niloticus*: behavioural and physiological responses [J]. Journal of the University of Kuwait Science, 15: 341-345.

ANDREWS GK, 2001. Cellular zinc sensors: MTF-1 regulation of gene expression [J]. Biometals, 14: 223-237.

ARIVAZHAGAN P, RAMANATHAN KPANNEERSELVAM C, 2001. Effect of D-Llipoic acid on mitochondrial enzymes in aged rats [J]. Chemico-Biol. Interact, 138: 189-198.

BAMBANG Y, THUET P, CHARMANTIER-DAURES M, et al, 1995. Effect of copper on survival and osmoregulation of various developmental stages of the shrimp penaeus-japonicus bate (Crustacea, Decapoda) [J]. Aquat. Toxicol, 33: 125-139.

BARTOSZ G, 2009. Reactive oxygen species: Destroyers or messengers? [J]. Biochem. Pharmacol, 77: 1303-1315.

BEYERS DW, FARMER MS, 2001. Effects of copper on olfaction of Colorado pikeminnow [J]. Environ. Toxicol. Chem, 20: 907-912.

BIEWENGA GP, HAENEN G, BAST A, 1997. The pharmacology of the antioxidant lipoic acid [J]. Gen. Pharmacol, 29: 315-331.

BIGOT A, DOYEN P, VASSEUR P, et al, 2009. Metallothionein coding sequence identification and seasonal mRNA expression of detoxification genes in the bivalve *Corbicula fluminea* [J]. Ecotoxicol. Environ. Saf, 72 (2): 382-387.

BOGDAN C, RöLLINGHOFF M, DIEFENBACH A, 2000. Reactive oxygen and reactive nitrogen intermediates in innate and specific immunity [J]. Curr. Opin. Immunol, 12: 64-76.

BOX A, SUREDA A, DEUDERO S, 2009. Antioxidant response of the bivalve *Pinna nobilis* colonised by invasive red macroalgae *Lophocladia lallemandii* [J]. Com. Biochem. Phys. C, 149: 456-460.

BREMNER I, 1998. Manifestations of copper excess [J]. Am. J. Clin. Nutr,

67: 1069S-1073S.

BRITTON RS, 1996. Metal-induced hepatotoxicity [J]. Semin. Liver Dis, 16: 3-12.

BRYA GW, 1984. Pollution due to heavy metals and their compounds [M] // KINNEO. A comprehensive integrate treatise of life in oceans and coastal waters. Wiley. Sons. Chichester, 1289-1431.

BUETTNER G, 1993. The packing order of free radicals and antioxidants: lipid peroxidation, alpha-tocopherol and ascorbate [J]. Arch. Biochem. Biophys, 300: 535-543.

ÇAKATAY U, 2006. Pro-oxidant actions of alpha-lipoic acid and dihydrolipoic acid [J]. Med. Hypotheses, 66: 110-117.

CAREY A, WILLIAMS, RHONDA M, et al, 2002. Lipoic acid as an antioxidant in mature thoroughbred geldings: A Preliminary Study [J]. American Society for Nutritional Sciences, 132: 1628S-1631S.

CEBALLOS-PICOT I, WITKO-SARSAT V, MERAD-BOUDIA M, et al, 1996. Glutathione antioxidant system as a marker of oxidative stress in chronic renal failure [J]. Free Radic. Biol. Med, 21: 845-853.

CHOW CK, 1979. Nutritonal influence on cellular antioxidant defense systems Amer [J]. Am. J. Cli. Nutr, 32 (5): 1066-1081.

CHOW CK, 1991. Vitamin E and oxidative stress [J]. Free Radical Biol. Med. 11: 215-232.

COMBS GF, COMBS SB, 1984. The nutritional biochemistry of selenium [J]. Ann. Rev. Nutr, 4: 257-280.

FANG YZ, YANG S, WU GY, 2002. Free Radicals, antioxidants, and nutrition [J]. Nutrition, 18: 872-879.

FRIBERG LED, 1979. Handbook on the Toxieology of Metals [M]. Amsterdam: Elsevier/North Holland Biochemical Press.

FRIDOVICH I, 1989. Superoxide dismutases. An adaptation to a paramagnetic gas [J]. J. Biol. Chem, 264: 7761-7764.

GERREKE PH, BIEWENGA, GUIDO R, et al, 1997. The pharmacology of the antioxidant lipoic acid [J]. Gen. Pharmac, 29: 315-331.

GUZMáN BP, GUTIéRREZ MS, MARCHETTI F, 2009. Methyl-parathion decreases sperm function and fertilization capacity targeting spermatocytes and maturing spermatozoa [J]. Toxicol. Appl. Pharm, 238: 141-149.

JI X, WANG W, CHENG J, et al, 2006. Free radicals and antioxidant status in rat liver after dietary exposure of environmental mercury [J]. Environ. Toxicol. Pharmacol, 22: 309-314.

JING G, LI Y, XIE L, 2006. Metalaccumulation and enzyme activities in gills and digestive gland of pearl oyster (*Pinctada fuca-ta*) exposed to copper [J]. Comp. Biochem. Physiol. C, 144 (2): 184-190.

KADIISKA MB, HANNA PM, JORDAN SJ, et al, 1993. Electron spin resonance evidence for free radical generation in copper-treated vitamin E-and selenium-deficient rats: in vivo spin-trapping investigation [J]. Mol. Pharmacol, 44: 222-227.

KASPERCZY KA, KASPERCZYK S, HORAK S, 2008. Assessment of semen function and lipid peroxidation among lead exposedmen [J]. Toxicol. Appl. Phama, 228: 378-384.

KLAUNIG JE, KAMENDULIS LM, 2004. The role of oxidative stress in carcinogenesis [J]. Annu. Rev. Pharmacol. Toxicol, 44: 239-267.

KOIZUMI T, LI ZG, TATSUMOTO H, 1992. DNAdamaging activity of cadmium in Leydig cells, a target cell population for cadmi-um carcinogenesis in the rat testis [J]. Toxicol. Lett, 63 (2): 211-220.

KOWLURU RA, ODENBACH S, BASAK S, 2005. Long-term administration of lipoic acid inhibits retinopathy in diabetic rats via regulating mitochondrial superoxide dismutase [J]. Invest Ophthalmol. Vis. Sci, 46: 396-422.

LANGMADE SJ, RAVINDRA R, DANIELS PJ, 2000. The transcription factor MTF-1 mediates metal regulation of the mouse ZnT1 gene [J]. J. Biol. Chem, 275: 34803-34809.

LAURIDSEN C, HOJSGAARD S, SORENSEN MT, 1999. Influence of dietary rapeseed oil, vitamin E, and copper on the performance and the antioxidative and oxidative status of pigs [J]. J. Anim. Sci, 77: 906-916.

LEONARD P, RYBAK, KAZIM HUSAIN, 1999. Dose dependent protection by lipoic acid against cisplatin-induced ototoxicity in rats: antioxidant defense system [J]. Toxicol. Sci, 47 (3): 195-202.

LESTER PACKER, 1995. In Proceedings of the International Symposium on Natural Antioxidants: Molecular Mechanism and Health Effects [M]. Accs.

Press.

LIMóN-PACHECO J, GONSEBATT ME, 2009. The role of antioxidants and antioxidant-related enzymes in protective responses to environmentally induced oxidative stress [J]. Mutat. Res, 674: 137-147.

LOPES PA, PINHEIRO T, SANTOS MC, 2001. Response of antioxidant enzymes in freshwater fish populations (Leuciscus alburnoides complex) to inorganic pollutants exposure [J]. Sci. Total Environ, 281 (1): 153-163.

MALEKI N, SAFAVI A, DOROODMAND MM, 2005. Determination of selenium in water and soil by hydride generation atomic absorption spectrometry using solid reagents [J]. Talanta, 66: 858-862.

MATTIE MD, FREEDMAN JH, 2001. Protective effects of aspirin and vitamin E (alpha-tocopherol) against copperand cadmium-induced toxicity [J]. Biochem. Biophys. Res. Commun, 285: 921-925.

MCGEER JC, SZEBEDINSZKY C, MCDONALD DG, et al, 2000. Effects of sublethal exposure to waterborne Cu, Cd or Zn in rainbow trout 1: ionoregulatory disturbance and metabolic costs [J]. Aquat. Toxicol, 50: 231-243.

MCPHERSON A, 1994. Selenium vitamin E and biological oxidation [M] // COLE DJ, GARNSWORTHY PJ. Recent advances in animal nutrition. Oxford: Butterworth and Heinemann's, 3-30.

MEISTER A, ANDERSON ME, 1983. Glutathione [J]. Annu. Rev. Biochem, 52: 711-760.

MIKHAILOVA M, LITTLEFIELD NA, HASS BS, 1997. Cadmium induced 8-hydroxydeoxyguanosine formation, DNA strand breaks and antioxidant enzyme activities in lymphoblastoid cells [J]. Cancer Lett, 115 (2): 141-148.

MOURENTE G, DIAZ-SALVAGO E, BELL JG, et al, 2002. Increased activities of hepatic antioxidant defence enzymes in juvenile gilthead sea bream (*Sparus aurata* L.) fed dietary oxidised oil: attenuation by dietary vitamin E [J]. Aquaculture, 214 (1): 343-361.

MURPHY R, DECOURSEY TE, 2006. Charge compensation during the phagocyte respiratory burst [J]. Biochim. Biophys. Acta, 1757 (8): 996-1011.

MYERS BM, PRENDERGAST FG, HOLMAN R, et al, 1993. Alterations in hepatocytes lysosomes in experimental hepatic copper overload in rats [J]. Gastroenterology, 105: 1814-1823.

NEWAIRY AA, EL-SHARAKY AS, BADRELDEEN MM, et al, 2007. The hepatoprotective effects of selenium against cadmium toxicity in rats [J]. Toxicology, 242: 23-30.

NORDBERG J, ARNER ES, 2001. Reactive oxygen species, antioxidants, and the mammalian thioredoxin system [J]. Free Radic. Biol. Med, 31 (11): 1287-1312.

OCHI T, ISHIGURO T, OHSAWA M, 1983. Participation of active oxygen species in the induction of DNA single-strand scissions by cadmium chloride in cultured Chinese hamster cells [J]. Mut. Res, 122 (2): 169-175.

OHHIRA M, ONO M, SEKIYA C, 1995. Changes in free radicalmetabolizing enzymes and lipid peroxides in the liver of Long _ /Evans with Cinnamon-like coat rats [J]. J. Gastroenterol, 30: 619-623.

OSSOLA JO, GROPPA MD, TOMARO ML, 1997. Relationship between oxidative stress and heme oxygenase induction by copper sulfate [J]. Arch. Biochem. Biophys, 337: 332-337.

PACKER L, KRAEMER K, RIMBACH G, 2001. Molecular aspects of lipoic acid in the prevention of diabetes complications. Nutrition, 17 (10): 888-895.

PACKER L, WITT E H, TRITSCHLER J, 1995. Alpha-lipoic acid as a biological antioxidant [J]. Free Radical Biol. Med, 19: 227-250.

PARK KH, TERJESEN BF, TESSER MB, et al, 2006. α-Lipoic acid-enrichment partially reverses tissue ascorbic acid depletion in pacu (*Piaractus mesopotamicus*) fed vitamin C-devoid diets [J]. Fish Physiol. Biochem, 32: 329-338.

POWELL SR, 2000. The antioxidant properties of zinc [J]. J. Nutr, 130: 1447S-1454S.

POWIS G, MONTFORT WR, 2001. Properties and biological activities of thioredoxins [J]. Annu. Rev. Pharmacol. Toxicol, 41: 261-295.

ROTRUER JT, POUE AL, 1993. Selenium: biochemical role as a component of glutathione peroxidase [J]. Science, 179: 588-589.

SANSINANEA AS, CERONE SI, STREITENBERGER SA, et al, 1998. Oxidative effect of hepatic copper overload [J]. Acta Physiol. Pharmacol. Ther. Latinoam, 48: 25-31.

SHARIFF M, JAYAWARDENA P, YUSOFF FM, et al, 2001. Immunological parameters of Javanese carp *Puntius gonionotus* (Bleeker) exposed to copper and challenged with *Aeromonas hydrophila* [J]. Fish shellfish immun, 11 (4): 281-291.

SHARMA G, NATH R, GILL KD, 1991. Effect of ethanol on cadmium-induced lipid peroxidation and antioxidant enzymes in rat liver [J]. Biochem. Pharmacol, 42 (S): 9-10.

SHUKLA GS, 1989. Cadmium toxicity and bioantioxidant: status of vitamin E and ascorbic acid of selected organs in rat [J]. J. Appl. Toxicol, 9 (2): 119-112.

SOKOL RJ, DEVEREAUX M, MIERAU G, et al, 1990. Oxidant injury to hepatic mitochondrial lipids in rats with dietary copper overload [J]. Gastroenterology, 90: 1061-1071.

SONG LS, ZOU HB, CHANG YQ, et al, 2006. The cDNA cloning and mRNA expression of a potential selenium-binding protein gene in the scallop Chlamys farreri [J]. Dev. Comp. Immunol, 30: 265-273.

STERN KG, 1936. On the mechanism of enzyme action a study of the decomposition of monoethyl hydrogen peroxide by catalase and of an intermediate enzyme-substrate compound [J]. J. Biol. Chem, 114 (2): 473-494.

TAKASHI, TAMURA AND THRESSA C, STADTMAN, 1996. A new selenoprotein from human lung adenocarcinoma cells: Prification, properties, and thioredoxin reductase activity [J]. Proc. Natl. Acad. Sci. USA, 93: 1006-1011.

THORNALLEY P, VASAK M, 1985. Possible for metallothionein in protection against radiation-in-duced oxidative stress kinetics and mechanism of reaction with superoxide and hydroxyl radicals [J]. Bioch. Biophys. Acta, 827: 36-44.

TRAN D, MOODY AJ, FISHER AS, et al, 2007. Protective effects of selenium on mercury-induced DNA damage in mussel haemocytes [J]. Aquat. Toxicol, 84: 11-18.

TREVISAN R, MELLO DF, FISHER AS, et al, 2011. Selenium in water enhances antioxidant defenses and protects against copper-induced DNA damage in the blue mussel Mytilusedulis [J]. Aquat. Toxicol, 101: 64-71.

VERLECAR XN, JENA KB, CHAINY GBN, 2008. Modulation of antioxidant defences in digestive gland of *Perna viridis* (L.), on mercury exposures [J].

Chemosphere, 71: 1977-1985.

WAOLKES MP, PERANTON A, BHAVE MR, et al, 1985. Interaetion of cadmium with insterstitial tissue of the rat testis: uptake of cadmium by isolated interstitial cells [J]. Biochem. Pharmaeo, 34 (14): 2513-2518.

YANG H, WANG X, LIU X, et al, 2009. Antioxidant peptidomics reveals novel skin antioxidant system. [J]. Mol. Cell. Proteomics, 8 (3): 571-583.

YEH ST, LIU CH, CHEN JC, 2004. Effect of copper sulfate on the immune response and susceptibility to Vibrio alginolyticus in the white shrimp Litopenaeus vannamei [J]. Fish Shellfish Immun, 17 (5): 437-446.

YOUSEF MI, KAMEL KI, GUENDIM IE, et al, 2007. Anin vitrostudy on reproductive toxicity of aluminium chloride on rabbit sperm: The protective role of some antioxidants [J]. Toxicology, 239: 213-223.

ZENZES MT, HONGMINN ZHANG, KRISHNAN S, et al, 1995. Cadmium accumulation in follicular fluid of women invitro fertilizationembryo transfer is higher in smokers [J]. Fertil-Steril, 64 (3): 599-603.

第二章
水中铜对皱纹盘鲍抗氧化反应、脂质过氧化和体内金属沉积的影响

第一节 引 言

在世界范围内,很多沿海地区,重金属是最常见的污染物,它们所带来的环境问题日益突出。重金属的毒性作用主要表现在可以破坏细胞膜或抑制细胞分裂,使水生动物生长和繁殖异常甚至死亡(邓道贵等,2002)。生物死亡最终会导致生态系统遭到破坏。在环境污染源中,重金属污染所带来的危害最为严重(刘家栋等,2001)。在养殖水域的消毒和控制疾病暴发方面,Cu 的使用都被证明为一种行之有效的方法,所以重金属 Cu 带来的污染广泛存在于养殖水体。重金属离子对水生生物的毒性作用之一是其可以造成机体内的氧化胁迫,在这个过程中会产生大量的活性氧自由基(ROS)。这些过量的 ROS 可使 DNA 断裂、酶蛋白失活和膜脂质过氧化(Bagchi 等,1996)等,因而导致机体的氧化应激,对机体诱发多种损害(Lemairon 等,1994;Stohs 等,2000;张迎梅等,2006)。

经过长期进化,为了对抗过氧化损害,需氧生物出现了抗氧化系统。抗氧化系统主要包括一些抗氧化的酶(SOD、CAT、GPx、GST 等)和一些抗氧化的小分子(GSH、维生素 E、维生素 C 等)。在正常的生理状态下,由代谢产生的过量的 ROS 可被抗氧化

防御系统所控制,通过调节体内一些抗氧化成分的改变,如 SOD、CAT 和 GPx 等,使机体免受氧化应激的伤害(刘慧等,2004;鲁双庆等,2002)。这些生物标志物的活性变化可以间接反映环境氧化应激的存在,因此常作为环境污染胁迫的敏感指标(Lemairon 等,1994)。

抗氧化系统和 ROS 之间的动态平衡是保证机体健康的重要因素,如果这种平衡被打破的话,就会出现氧化应激,从而对生物体产生危害(Sies,2000)。对水生动物来说这种平衡更容易被打破。海水贝类很容易受到环境污染物的影响,因为贝类具有聚集这些污染物的特性,从而加强了 ROS 的产生。

皱纹盘鲍(*Haliotis discus hannai* Ino)是海洋大型原始腹足类,是中国海水养殖贝类的主要品种之一。本章重点研究不同梯度的铜对皱纹盘鲍的毒性作用。抗氧化防御系统包括抗氧化酶和抗氧化的小分子 GSH。以脂质过氧化生物指标为依据,对皱纹盘鲍进行生物监测,为防治水体重金属污染提供科学依据。这样既可以加强鲍的养殖管理,又可以提高水产动物的食物安全性,为找到合适的解毒途径提供条件。

第二节 摄食生长实验的设计

(一)实验动物与药品

皱纹盘鲍幼鲍和新鲜的海带均购买于青岛鳌山卫育苗场,皱纹盘鲍为同一批鲍苗。在中国海洋大学水产馆养殖系统中暂养 2 周。暂养期间,每天 18:00 饱食投喂海带 1 次,次日 8:00 清底换水,养殖过程中密切观察采食及健康状况。$CuSO_4 \cdot 5H_2O$(AR)购自广州化学试剂厂。实验前,先用蒸馏水配制成 Cu 浓度为 1 000mg/L 的母液。再根据实验需要,把母液在 5 个 350L 实验用洁净桶内用天然海水稀释成实验所需要的各种浓度,然后用水泵抽到各个玻璃缸中,玻璃缸的体积为 100L。硝酸、硫酸、过氧化氢、盐酸均为优级纯。

(二) 实验设计

根据直线内插法得到的重金属 Cu 对皱纹盘鲍的半致死浓度为 0.164mg/L，且累计死亡率（y）和浓度（x）的相关直线方程为：$y=15.83x-2.088$（$R^2=0.956$）。根据 Cu 的半致死浓度和国家《渔业水质标准》（GB 11607—1989）（Cu≤0.01mg/L）确定铜的亚慢性毒性实验浓度。实验设 0（对照组）、0.02mg/L、0.04mg/L、0.06mg/L、0.08mg/L 5 个处理组，每个处理组均设 3 个重复。海水中 Cu 的实测浓度为 4.5μg/L、（0.024±0.00）mg/L、（0.045±0.00）mg/L、（0.065±0.00）mg/L、（0.082±0.00）mg/L。挑选规格一致的健康个体随机分成 15 组，每缸 60 只鲍。在整个实验过程中，保持水中溶氧量在 6mg/L 以上，pH 为 7.4～7.9，水温控制在 18～22℃，盐度为 2.2～2.8，光暗比为 12h∶12h。实验时间为 28d，每天早晚换水 2 次，更换总量为各组溶液的 50%。每天 18∶00 投喂新鲜海带 1 次，第二天 8∶00 清除粪便和残饵。如有个体死亡，即时捞出并且计数。实验海水的水样每 2d 取一次以便以后实验分析。实验检测海带中 Cu 的含量为 4.45mg/kg。

(三) 取样与样品处理

于暴露的第 0、1、3、6、10、15、21、28 天取样，每次从各个暴露组中取 8 只活的鲍，并在实验的第 28 天将所有活的鲍全部取样。取样的鲍不归入死亡率的计算。对所有的鲍在冰盘中取出肝脏组织和肌肉组织，并将每组的组织剪成小块，混匀后分装在离心管中，放置在 -80℃ 冰箱中待测。

第三节　铜浓度对皱纹盘鲍死亡率和抗氧化反应的影响

(一) 对死亡率的影响

如图 2-1 所示，对照组和 0.02mg/L Cu 浓度的实验组鲍在实验的过程中没有出现死亡，但是 0.04、0.06、0.08mg/L Cu 浓度的实验组鲍分别从实验的第 7、5、2 天开始出现死亡，并且分别在实验的第 21、10、6 天全部死亡。

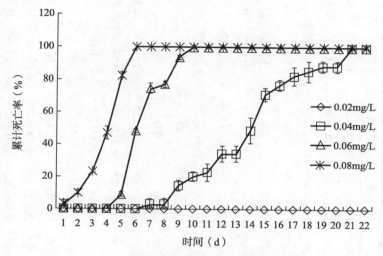

图 2-1　水体中的铜对皱纹盘鲍累计死亡率的影响

（二）对抗氧化反应的影响

如表 2-1 所示，总体上来看，各 Cu 浓度下肝胰脏 SOD 的活力随着时间的增长呈下降的趋势。在实验的第 1 天，0.04、0.06、0.08mg/L Cu 浓度的实验组显著低于对照组。总体上来看，从实验的第 1~21 天，各 Cu 浓度下 CAT 的活力都高于对照组。在实验的第 1、3 天，各 Cu 浓度下的 CAT 活力显著高于对照组。在 0.02、0.04、0.08mg/L Cu 浓度下，CAT 的活力最高值均出现在实验的第 1 天。在 0.06mg/L Cu 浓度下，CAT 的活力最高值出现在实验的第 3 天。在各 Cu 浓度下，GPx 的活力在实验的第 1、3、6、10、15 天都高于对照组。在 0.06mg/L Cu 浓度下，肝胰脏中 GPx 的活力在实验的第 3、10 天均显著高于对照组。在 0.08mg/L Cu 浓度下，GPx 的活力在实验的第 6 天显著高于对照组。在 0.02、0.04mg/L Cu 浓度下，GPx 的活力分别在实验的第 1~15 天和第 1~21 天均显著高于对照组。总体上看，在各 Cu 浓度下，GST 的活力随着时间延长呈

表 2-1 水体中 Cu 对皱纹盘鲍肝胰脏抗氧化相关指标的影响（平均值±标准误，$n=3$）

指标	Cu浓度	第0天	第1天	第3天	第6天	第10天	第15天	第21天	第28天
SOD	对照	29.66±0.55	30.12±1.26^A	28.74±0.39^A	28.53±0.44^A	27.82±1.15^A	27.82±1.15^A	27.38±1.36^A	29.69±0.82^A
	0.02mg/L	29.66±0.55	33.74±1.96^A	21.92±1.18^{BC}	29.52±0.71^A	19.38±1.68^B	19.38±1.68^B	19.34±1.01^B	17.63±0.81^B
	0.04mg/L	29.66±0.55^a	20.03±1.68^{Bab}	17.73±0.18^{Cab}	26.09±0.40^{Bab}	20.17±0.17^{Bab}	20.17±1.69^{Bab}	16.53±0.18^{Bb}	ND
	0.06mg/L	29.66±0.55^a	19.82±1.79^{Bb}	24.87±1.69^{ABab}	18.24±0.08^{Cab}	20.17±1.46^{Bab}	ND	ND	ND
	0.08mg/L	29.66±0.55^a	16.81±1.59^{Bb}	19.97±0.84^{Cb}	19.10±0.68^{Cab}	ND	ND	ND	ND
CAT	对照	10.78±0.40	10.81±0.66^B	10.66±1.18^B	11.95±0.13^B	10.47±0.87^B	9.52±0.25^B	9.86±0.30^B	11.33±0.82
	0.02mg/L	10.78±0.40^c	19.60±2.32^{Aa}	18.90±1.32^{Aa}	17.08±0.83^{Aab}	15.10±0.08^{Aab}	14.68±1.26^{Aab}	13.61±1.33^{ABabc}	9.02±1.71^c
	0.04mg/L	10.78±0.40^c	20.04±0.42^{Aa}	16.50±1.27^{Aab}	16.98±0.88^{Aab}	12.79±1.37^{Aab}	16.82±1.52^{Aab}	13.91±0.68^{Aab}	ND
	0.06mg/L	10.78±0.40^b	17.44±0.29^{Aa}	19.77±1.01^{Aa}	12.08±0.25^{Bb}	12.61±0.84^{Bb}	ND	ND	ND
	0.08mg/L	10.78±0.40^b	20.35±1.03^{Aa}	18.66±0.40^{Aa}	11.07±1.14^{Bb}	ND	ND	ND	ND
GPx	对照	50.75±2.58	53.11±2.67	46.46±4.60^B	47.95±1.67^B	48.66±2.63^B	49.05±1.27^C	58.7±4.56^B	52.13±4.19^A
	0.02mg/L	50.75±2.58^c	58.75±5.73^{abc}	55.59±4.12^{ABabc}	69.20±4.41^{Aab}	74.24±8.55^{Aa}	69.85±1.42^{Aab}	41.38±0.43^{Cac}	25.83±2.94^{dB}
	0.04mg/L	50.75±2.58^c	70.14±11.25^{bc}	55.59±1.44^{ABc}	62.82±2.15^{Abc}	86.53±6.18^{Aa}	114.86±7.45^{Aa}	70.96±1.26^{Aa}	ND
	0.06mg/L	50.75±2.58^c	74.76±6.91^b	64.14±1.06^{Aab}	62.80±1.07^{Aab}	74.70±2.48^{Aa}	ND	ND	ND
	0.08mg/L	50.75±2.58^b	77.70±6.03^b	60.23±1.49^{Aab}	62.37±2.92^{Aa}	ND	ND	ND	ND
GST	对照	48.38±0.55	57.01±3.37	51.63±1.18^A	48.73±1.99^{AB}	57.87±1.84^A	56.24±2.37^A	51.98±4.10^A	53.13±1.76^A
	0.02mg/L	48.38±0.55^{ab}	51.78±1.30^a	56.34±2.13^{Aa}	55.83±6.42^{Aa}	51.58±3.00^{AB}	62.24±1.39^{Aa}	35.28±1.96^{Bbc}	27.98±1.63^{Bc}

(续)

指标	Cu 浓度	第0天	第1天	第3天	第6天	第10天	第15天	第21天	第28天
GST	0.04mg/L	48.38±0.55^{ab}	47.67±1.73^{abc}	57.18±4.38^{Aa}	25.14±2.14^{Cd}	33.02±4.78^{Bd}	33.80±0.02^{Bcd}	36.36±2.96^{Bcd}	ND
	0.06mg/L	48.38±0.55^{ab}	44.83±2.01^{b}	56.61±5.43^{Aa}	22.32±1.57^{Cc}	29.78±2.17^{Bc}	ND	ND	ND
	0.08mg/L	48.38±0.55^{ab}	56.81±4.97^{a}	33.55±6.13^{Bb}	33.38±4.73^{Bb}	ND	ND	ND	ND
GSH	对照	4.47±0.04	4.07±0.20^{B}	4.19±0.27^{B}	4.23±0.08^{B}	3.92±0.19^{B}	3.50±0.45^{B}	3.90±0.32^{B}	4.37±0.32
	0.02mg/L	4.47±0.04^{c}	7.11±0.37^{Aabc}	7.23±0.67^{Aa}	7.55±0.93^{Aa}	7.78±0.69^{Aa}	7.27±0.46^{Aab}	5.38±0.29^{Abc}	4.80±0.34^{bc}
	0.04mg/L	4.47±0.04^{c}	6.44±0.36^{Abcd}	5.64±0.39^{ABcde}	5.49±0.23^{ABe}	5.95±0.46^{ABc}	8.05±0.25^{Aa}	7.75±0.77^{Aab}	ND
	0.06mg/L	4.47±0.04^{a}	6.85±0.21^{Aa}	7.35±0.47^{Aa}	5.39±0.36^{ABa}	6.74±0.21^{Aa}	ND	ND	ND
	0.08mg/L	4.47±0.04^{b}	7.44±0.71^{Aa}	6.18±0.44^{ABa}	6.54±0.84^{ABab}	ND	ND	ND	ND
T-AOC	对照	2.84±0.07^{a}	2.32±0.09	2.39±0.12^{A}	2.43±0.19^{A}	2.29±0.15^{A}	2.55±0.02^{A}	2.30±0.18^{A}	2.73±0.17^{A}
	0.02mg/L	2.84±0.08^{a}	2.53±0.07^{a}	2.38±0.14^{ab}	2.41±0.17^{Aa}	1.42±0.15^{Ba}	1.81±0.07^{Bc}	1.81±0.10^{ABc}	1.60±0.12^{Bc}
	0.04mg/L	2.84±0.09^{a}	2.46±0.17^{ab}	2.27±0.05^{ABbc}	1.77±0.10^{Bcd}	1.63±0.11^{Bd}	1.94±0.13^{Bcd}	1.75±0.07^{Bcd}	ND
	0.06mg/L	2.84±0.10^{a}	2.37±0.17^{a}	2.29±0.13^{ABa}	1.51±0.13^{Bb}	1.58±0.12^{Bb}	ND	ND	ND
	0.08mg/L	2.84±0.10^{a}	2.28±0.16^{b}	1.82±0.02^{Bc}	1.74±0.03^{Bc}	ND	ND	ND	ND

注：CAT，过氧化氢酶（U，每毫克蛋白中）；SOD，超氧化物歧化酶（U，每毫克蛋白中）；GPx，谷胱甘肽过氧化物酶（U，每毫克蛋白中）；GST，谷胱甘肽硫转移酶（U，每毫克蛋白中）；GSH，还原性谷胱甘肽（mg，每克蛋白中）。同行数据中，经 Tukey 检验差异不显著的平均值之间用相同的大写字母表示（$P>0.05$）；同列数据中，经 Tukey 检验差异不显著的平均值之间用相同的小写字母表示（$P>0.05$）。

ND：无数据。

下降的趋势。在 0.02mg/L Cu 浓度下，GST 的活力在实验的第 21、28 天显著降低。在 0.04、0.06、0.08mg/L Cu 浓度下，肝胰脏 GST 的活力分别从实验的第 6、6、3 天显著降低。在各 Cu 浓度下，GSH 的含量均高于对照组，并且在实验的第 1、10、15 天均显著高于对照组。在 0.06mg/L Cu 浓度下分别在实验的第 1、3、10 天显著高于对照组。在各 Cu 浓度下，各实验组肝胰脏的总抗氧化力（T-AOC）水平呈现下降的趋势。在 0.02mg/L Cu 浓度下，T-AOC 从实验的第 10 天开始显著性下降。在 0.04、0.06、0.08mg/L Cu 离子浓度下，T-AOC 分别从实验的第 6 天、第 6 天和第 3 天显著下降。

大量研究表明（Company 等，2004；Tamás 等，2009），Cu 离子的氧化胁迫可以导致机体过量活性氧（ROS）的形成，从而导致机体的氧化应激。在维持氧自由基平衡方面生物体内抗氧化系统起着非常重要的作用（Palacea 等，1998）。SOD、CAT、GPx、GST 和 GSH 是贝类内抗氧化酶的重要组分。在正常生理情况下，机体内的抗氧化酶系统可以有效地清除体内的超氧阴离子（O^{2-}）、单线态氧（$^1O_2$）、羟自由基（·OH）和 H_2O_2 等活性氧物质，维持氧化和抗氧化的平衡，保护机体免受自由基伤害（鲁双庆等，2002）。在本实验 Cu 离子的胁迫下，抗氧化酶和抗氧化的小分子 GSH 均呈现显著性变化，这可能是一个机体的抵抗机制，表明 Cu 离子胁迫下皱纹盘鲍肝胰脏 ROS 的增加。以前的研究和本研究均证明，这些抗氧化酶和抗氧化的小分子 GSH 在抵抗 Cu 导致的氧化应激中发挥着重要的作用，所以它们经常作为污染物的早期检测的指标。本研究发现，在 Cu 离子的胁迫下，肝胰脏的 CAT、GPx 的活力和 GSH 的含量随着时间的延长显著性升高，而 SOD 和 GST 的活力显著性降低，而且 SOD、CAT 的活力及 GSH 的含量在实验的第 1 天就发生了显著性变化，GST 和 GPx 的活力在实验的第 3 天也发生显著性变化。这说明 CAT、GPx 和 GSH 在皱纹盘鲍抵抗 Cu 的氧化应激中发挥了重要的作用，且 CAT 和 GSH 在抵抗 Cu 离子过程中更加敏感。

第二章 水中铜对皱纹盘鲍抗氧化反应、脂质过氧化和体内金属沉积的影响

SOD 是金属酶，含 Cu、Zn 或 Mn 原子，SOD 可以催化超氧阴离子产生 H_2O_2 和水，抑制了超氧阴离子向·OH 的转化（Kappus，1985；Ruas 等，2008）。CAT 是过氧化物酶的标志性酶，可以催化 H_2O_2 转化成无害物质水和氧气。GPx 主要清除有机氢过氧化物和脂类的氢过氧化物，并且在 CAT 含量少或 H_2O_2 低的组织中，可以替代 CAT 清除 H_2O_2。本研究发现，皱纹盘鲍肝胰脏 SOD 的活力随着 Cu 离子浓度的增加和时间的增长呈下降的趋势。在 $25.6\mu g/L$ Cu 离子胁迫下，贻贝（*Bathymodiolus azoricus*）鳃的 SOD 活力在 24h 内一直被抑制（Company 等，2004）。Jiang 等（2011）报道，在 $0.6 \sim 7.2 mg/L$ Cu 离子胁迫下，鲤（*Cyprinus carpio*）肝胰脏 SOD 的活力显著性被抑制。然而，本实验中在各 Cu 离子浓度的胁迫下，肝胰脏 CAT 的活力先升高后降低到对照水平，GPx 的活力在各个 Cu 浓度下也是显著性高于对照组。这表明皱纹盘鲍肝胰脏 CAT 和 GPx 活力的升高是对 SOD 活力被抑制的补偿反应。从另一个角度来看，CAT 活力的变化表明 CAT 正逐渐失去对抗自由基的能力。

在各 Cu 离子浓度下，皱纹盘鲍肝胰脏 GST 的活性随时间延长呈下降的趋势，且与对照组有显著性差异。根据 Cunha 等（2007）报道，在 Cu 离子胁迫下，海水腹足类（*Nucella lapillus*）GST 的活力降低了 47.6%。在 100、$250\mu g/L$ Cu 离子胁迫下，鲤的 GST 活力也被显著性抑制（Dautrememepuits 等，2002）。GST 可以催化某些外来有害物质或内源性亲电子基团与 GSH 的巯基偶联，增加其疏水性，使其容易穿越细胞膜，分解排出体外。实验中 Cu 离子胁迫下，GST 的活力被抑制的原因可能是过多的 ROS 抑制了肝胰脏 GST 的活力。

GSH 在生物体内的解毒中起重要的作用，主要表现在两个方面：①GSH 作为 GPx 和 GST 的底物，可以通过这两种抗氧化酶起解毒作用。②GSH 可以直接与有机体的氧自由基及一些亲电化合物结合起解毒作用（Stegeman 等，1992）。根据 Al-Subiai 等（2009）研究发现，在 $40\mu g/L$ Cu 离子胁迫下，紫贻贝（*Mytilus edulis*）

肌肉 GSH 的含量显著性升高。在本实验各 Cu 离子的胁迫下，肝胰脏中的 GSH 被诱导，表明肝胰脏作为皱纹盘鲍重要的代谢和解毒器官，其中的 GSH 活跃地参与 Cu 离子的代谢活动。冯涛等（2001）研究认为，肝胰脏 GSH 含量的升高被认为是机体对污染物暴露的一种适应性反应。

总抗氧化力（T-AOC）是近年用于衡量生物体抗氧化系统（包括酶系统和非酶促系统）功能状况的综合性指标，可以反映生物体抗氧化系统对外来刺激的代偿能力以及机体自由基的代谢状态，其中的酶系统主要包括 SOD、CAT、GPx、GST 等；非酶系统主要包括 GSH，维生素 E、维生素 C、维生素 K、维生素 A、尿酸、血浆铜蓝蛋白、胡萝卜素、葡萄糖等（Mukai 等，1993；Burton 等，1990）。本研究中，皱纹盘鲍暴露在各浓度 Cu 离子溶液中时，CAT、GPx 的活力及 GSH 的含量随时间延长呈上升的趋势，但是 SOD 和 GST 的活力呈下降的趋势。虽然不同的抗氧化成分对 Cu 离子的氧化应激的反应不同，但是 T-AOC 呈下降的趋势说明在 Cu 离子胁迫下，皱纹盘鲍肝胰脏抗氧化力水平随着时间的延长而下降。

第四节 铜浓度对皱纹盘鲍脂质过氧化和体内金属沉积的影响

（一）对脂质过氧化的影响

如表 2-2 所示，在 0.02mg/L Cu 浓度下，肝胰脏丙二醛（MDA）的含量随时间延长呈先升高后下降的趋势。然而高浓度 Cu 离子下，MDA 的含量随时间延长呈升高的趋势。在 0.02mg/L Cu 浓度下，MDA 的含量在实验的第 3 天显著升高，从实验的第 10 天开始下降，到第 28 天降低到对照水平。在 0.04mg/L Cu 浓度下，MDA 的含量在实验的第 6 天显著高于对照组。在 0.06mg/L 和 0.08mg/L Cu 离子浓度下，MDA 的含量在整个暴露实验中均高于对照组。

表 2-2 水体中 Cu 对皱纹盘鲍肝胰脏 MDA 含量的影响（平均值±标准误，$n=3$）

Cu 浓度	第 0 天	第 1 天	第 3 天	第 6 天	第 10 天	第 15 天	第 21 天	第 28 天
对照	4.93±0.12	5.23±0.20	5.56±0.46^B	5.23±0.20^C	4.62±0.26^{AB}	5.58±0.11^{AB}	4.73±0.40^{AB}	6.23±0.38
0.02mg/L	4.93±0.12^{bc}	6.59±0.65^{bc}	9.37±0.70^{Aa}	6.59±0.65^{ABab}	6.40±0.33^{ABc}	4.78±0.56^{Bc}	4.44±0.44^{Bc}	4.96±0.67^{bc}
0.04mg/L	4.93±0.12^{cd}	6.38±0.09^{bc}	6.60±0.68^{Bbc}	8.70±0.02^{Aa}	4.47±0.56^{Bd}	7.01±0.24^{Aab}	5.97±0.08^{Abcd}	ND
0.06mg/L	4.93±0.12^b	5.91±0.03^b	6.83±0.54^{Ba}	5.91±0.03^{BCab}	5.95±0.40^{ABab}	ND	ND	ND
0.08mg/L	4.93±0.12^b	6.21±0.54^{ab}	7.16±0.17^{ABa}	6.74±0.12^{BCa}	ND	ND	ND	ND

注：MDA，丙二醛（nmol，每毫克蛋白中）。
同行数据中，经 Tukey 检验差异不显著的平均值之间用相同的小写字母表示（$P>0.05$）；同列数据中，经 Tukey 检验差异不显著的平均值之间用相同的大写字母表示（$P>0.05$）。
ND：无数据。

丙二醛（MDA）是脂质过氧化的中间产物（Ohkawa等，1979），可以对膜结构造成严重损伤。Cu可以通过Fenton氧化还原循环反应产生ROS（Halliwell和Gutteridge，1984），并且可以参与脂质过氧化的激活和连锁反应（Viarengo等，1990）。在本实验中，0.02mg/L Cu离子实验组MDA含量在实验的第6天被显著性激活，且其他Cu离子浓度的实验组MDA含量在第6天均高于对照组。很多研究发现，Cu离子胁迫可以导致水产动物脂质过氧化的发生，例如导致河口鱼类 *Pomatoschistus microps* 肝脏的脂质过氧化增加（Vieira等，2009）。根据Chelomin和Belcheva（1991），随着扇贝 *Mizuhopecten yessoensis* 肝胰脏Cu积累量的升高，MDA的含量显著性增加。在 $100\mu g/L$ Cu离子浓度暴露7d，淡水蟹（*Oziotelphusa senex senex*）MDA的含量显著性升高（Reddy和Bhagyalakshmi，1994）。然而在本实验中，在低浓度0.02mg/L Cu离子浓度作用下，肝胰脏MDA的含量随时间延长先升高后下降到对照水平。推测可能是机体抗氧化系统发挥了作用，水中Cu离子含量超过0.04mg/L对于皱纹盘鲍可能过高，从而导致脂质过氧化的发生。

（二）对体内金属沉积的影响

如表2-3所示，总体上来说，肝胰脏和肌肉的Cu含量随着水中Cu离子浓度的升高和时间的延长呈上升的趋势。肝胰脏、肌肉中Cu离子含量最高值为81.56、$24.57\mu g/g$，分别是对照组的10、5倍。

软体动物可以积累高浓度的重金属，并作为海洋重金属污染的指标生物（Langston等，1998）。本实验研究表明，Cu离子的含量随Cu离子暴露时间的延长和离子浓度的增加而呈上升的趋势。由于鲍的死亡，0.04、0.06、0.08mg/L的Cu浓度实验组分别在实验的第21、10、6天结束。然而，0.06mg/L Cu浓度第10天和0.04mg/L Cu浓度第21天实验组的肝胰脏和肌肉中Cu的积累量高于0.08mg/L Cu浓度第6天的积累量。换句话说，Cu积累速率较慢的实验组在较长时间里积累的Cu含量较高，组织Cu的含量并没有积累到一个导致皱纹盘鲍死亡的临界值。本实验得出，组织中Cu的积累量和死亡率无显著相关性。

第二章 水中铜对皱纹盘鲍抗氧化反应、脂质过氧化和体内金属沉积的影响

表2-3 水体中Cu对皱纹盘鲍肝胰脏和肌肉中Cu含量（$\mu g/g$）的影响（平均值±标准误，$n=3$）

组织	Cu浓度	第0天	第1天	第3天	第6天	第10天	第15天	第21天	第28天
肝胰脏	对照	8.09±0.63	8.40±0.86	9.31±0.01^C	8.51±0.34^C	8.65±0.44^{Cd}	7.77±0.81^C	8.31±0.43^C	7.72±0.51^B
	0.02mg/L	8.09±0.63^e	11.98±0.96^{de}	14.33±0.95^{BCcd}	15.71±1.50^{BCcd}	20.27±0.54^{BCc}	24.20±1.29^{Bb}	24.63±1.94^{Bb}	38.12±1.42^{Aa}
	0.04mg/L	8.09±0.63^f	14.37±1.99^{ef}	18.77±1.24^{ABe}	31.56±1.62^{ABcd}	39.03±1.18^{ABc}	68.38±1.74^{Ab}	81.56±1.58^{Aa}	ND
	0.06mg/L	8.09±0.63^e	12.49±1.41^{bc}	24.63±2.26^{Abc}	30.66±9.55^{ABb}	66.97±3.77^{Aa}	ND	ND	ND
	0.08mg/L	8.09±0.63^e	18.73±4.54^{bc}	24.56±2.90^{ABb}	39.02±2.48^{Aa}	ND	ND	ND	ND
肌肉	对照	5.88±0.60	5.29±0.25^B	5.84±0.20^B	5.14±0.37^B	5.64±0.21^C	4.93±0.34^C	4.74±0.46^B	5.47±0.37^B
	0.02mg/L	5.88±0.60^d	7.97±0.70^{ABCd}	9.77±1.53^{ABbcd}	8.39±0.70^{Bcd}	13.14±0.94^{Bb}	10.73±0.84^{Bbc}	12.62±0.43^{Bbc}	14.96±0.90^{Aa}
	0.04mg/L	5.88±0.60^d	8.96±0.21^{Ad}	13.95±0.86^{Acd}	13.67±0.75^{Abcd}	20.74±1.21^{Aab}	17.08±0.72^{Aabc}	24.57±3.17^{Aa}	ND
	0.06mg/L	5.88±0.60^c	6.25±0.50^{ABc}	13.93±0.28^{Ab}	15.67±0.61^{Aab}	21.63±0.73^{Aa}	ND	ND	ND
	0.08mg/L	5.88±0.60^c	8.51±0.95^{Abc}	13.32±1.06^{Aab}	18.18±2.09^{Aa}	ND	ND	ND	ND

注：同行数据中，经Tukey检验差异不显著的平均值之间用相同的大写字母表示（$P>0.05$）；同列数据中，经Tukey检验差异不显著的平均值之间用相同的小写字母表示（$P>0.05$）。ND：无数据。

在本实验中，在 0.04、0.06、0.08mg/L Cu 浓度下（它们分别是国家水质标准的 4、6、8 倍），分别在实验的第 21、10、6 天导致了皱纹盘鲍的全部死亡。Wang 等（2007a，2007b）研究表明，软体动物对 Cu 的敏感性要高于黑头鱼（*Pimephales promelas*）和虹鳟（*Oncorhynchus mykiss*）。在 2、12μg/L Cu 离子浓度下，贻贝死亡率显著升高，分别为对照组的 20.9% 和 69.9%（Jorge 等，2013）。在实验各 Cu 离子浓度下，抗氧化酶和抗氧化小分子 GSH 在暴露过程中显著性升高或显著性降低，且肝胰脏脂质过氧化的水平显著性升高。这说明 Cu 导致的氧化应激导致了皱纹盘鲍抗氧化防御系统的损伤，从而导致了机体的脂质过氧化或鲍的死亡。

小结：

（1）在不同浓度 Cu 离子作用下皱纹盘鲍肝胰脏 CAT、GPx 活性和 GSH 被诱导，SOD 和 GST 被显著性抑制。但是 T-AOC 随浓度和时间的延长呈下降的趋势，而且皱纹盘鲍肝胰脏 MDA 含量显著升高。故猜测是这两个方面导致皱纹盘鲍的死亡。

（2）皱纹盘鲍肝胰脏 SOD、CAT、GPx、GST 活性和 T-AOC 对 Cu 的早期污染有指示作用，其中最为灵敏的指标是 CAT 和 GSH。

参 考 文 献

邓道贵，张桂凤，耿雪侠，2002. Cu^{2+} 对日本沼虾幼虾的急性致毒研究 [J]. 淮北煤炭师学院学报，23（3）：36-38.

冯涛，郑微云，陈荣，2001. 苯并（α）芘对大弹涂鱼肝脏和卵巢还原型谷胱甘肽含量影响的比较研究 [J]. 海洋环境科学，20（1）：12-15.

刘慧，王晓蓉，张景飞，等，2004. 铜及其 EDTA 配合物对彭泽鲫鱼肝脏抗氧化系统的影响 [J]. 环境化学，23（3）：263-267.

刘家栋，翟兴礼，王东平，2001. 植物抗盐机理的研究 [J]. 农业与技术，21（1）：26-29.

鲁双庆，刘少军，刘红玉，等，2002. Cu 对黄鳝肝脏保护酶 SOD、CAT、GSH-PX 活性的影响 [J]. 中国水产科学，9（2）：138-141.

张迎梅，王叶菁，虞闰六，等，2006. 重金属 Cd^{2+}、Pb^{2+} 和 Zn^{2+} 对泥鳅 DNA

损伤的研究 [J]. 水生生物学报, 30 (4): 399-403.

AL-SUBIAI SN, JHA AN, MOODY AJ, et al, 2009. Contamination of bivalve haemolymph samples by adductor muscle components: implications for biomarker studies [J]. Ecotoxicol, 18: 334-342.

BAGCHI D, BACGHI M, HASSOUN EA, et al, 1996. Cadmium induced excretion of urinary lipid metabolites, DNA damage, glutathione depletion and hepatic lipid peroxidation in Sprague-Dawley rats [J]. Biol. Trace Elem. Res, 52: 143-154.

BURTON G W, TRABER MG, 1990. Vitamin E: antioxidant activity biokinetics and bioavailability [J]. Annu. Rev. Nutr, 120: 357-382.

CHELOMIN VP, BELCHEVA NN, 1991. Alterations of microsomal lipid synthesis in gill cells of bivalve mollusk *Mizuhopecten yessoensisin* response to cadmium accumulation [J]. Comp. Biochem. Physiol (C), 99: 1-5.

COMPANY R, SERAFIM A, BEBIANNO MJ, et al, 2004. Effect of cadmium, copper and mercury on antioxidant enzyme activities and lipid peroxidation in the gills of the hydrothermal vent mussel *Bathymodiolus azoricus* [J]. Mar. Environ. Res, 58: 377-381.

CUNHA I, MANGAS-RAMIREZ E, GUILHERMINO L, 2007. Effects of copper and cadmium on cholinesterase and glutathione S-transferase activities of two marine gastropods (Monodonta lineata and Nucella lapillus)[J]. Comp. Biochem. Physiol (C), 145: 648-657.

DAUTREMEMEPUITS C, BETOULLE S, VERNET G, 2002. Antioxidant response modulated by copper in healthy and parasitized carp (*Cyprinus carpio* L.) by Ptychobothrium sp. (Cestoda) [J]. Biochim. Biophys. Acta, 1573: 4-8.

HALLIWELL B, GUTTERIDGE MC, 1984. Oxygen toxicity, oxygen radicals, transition metals and disease [J]. Biochem. J, 219: 1-14.

JIANG WD, WU P, KUANG SY, 2011. Myo-inositol prevents copper-induced oxidative damage and changes in antioxidant capacity in various organs and the enterocytes of juvenile Jian carp (*Cyprinus carpio var.* Jian) [J]. Aquat. Toxicol, 105: 543-551.

JORGE MB, LORO VL, BIANCHINI A, et al, 2013. Mortality, bioaccumulation and physiological responses in juvenile freshwater mussels (*Lampsilis siliquoidea*) chronically exposed to copper [J]. Aquat. Toxicol, 126: 137-147.

KAPPUS H, 1985. Lipid peroxidation: mechanisms, analysis, enzymology and biological relevance. In: Sies H (ed) Oxidative Stress [M]. London: Academic. Press, 40 (3): 273-310.

LANGSTON WJ, BEBIANNO MJ, BURT GR, 1998. Metal handling strategies in molluscs. In: Langston WJ, Bebianno MJ (eds) Metal Metabolism in Aquatic Environments [M]. London: Chapman and Hall.

LEMAIRON P, MATTEWS A, FORLIN L, 1994. Stimulation ofoxyradical production ofhepaticmicrosomes of flounder (Platichthysfle-sus) and perch (Perca fluviatilis) bymodeland pollutantxenobi-otics [J]. Arch Environ. Contour. Toxicoh, 26: 191-200.

MUKAI K, MORINCOTO H, OKAUCHI Y, 1993. Kinetic study of reactions between tocopheroxyl radicals and fatty acids [J]. Lipid, 28: 753-756.

OHKAWA H, OHISH IN, YAGI K, 1979. Assay for lipid peroxidation in animal tissues by thiobarbituric acid reaction [J]. Anal. Biochem, 95: 351-363.

PALACEA VP, BROWN SB, BARON CI, 1998. An evaluation of the relationships among oxidative stress, antioxidant vitam ins and early mortality syndrome (EMS) of lake trout (*Salvelinus namaycush*) from Lake Ontario [J]. Aquat. Toxicol, 43: 259-268.

REDDY PS, BHAGYALAKSHMI A, 1994. Lipid peroxidation in the gill and hepatopancreas of *Oziotelphusa senex senex* fabricius during cadmium and copper exposure [J]. Bull. Enviro. Contam. Toxicol, 53: 704-710.

RUAS CBG, CARVALHO CS, DE ARAUJO HSS, et al, 2008. Oxidative stress biomarkers of exposure in the blood of cichlid species from a metal-contaminated river [J]. Ecotoxicol. Environ. Saf, 71: 86-93.

SIES H, 2000. Oxidative stress [J]. In: Encyclopedia of Stress , 3: 102-105.

STEGEMAN JJ, BROUWER M, DIGIULIO RT, 1992. Molecular responses to environmental contamination: enzyme and pr otein synthesis as indicators of chemical exposure and effects. Huggett RJ, Kimerle RA, Mehrle PM. Biomarkers, biochemical, physiological and histological markers of anthropogenic stress [M]. Boca. Raton. Florida: Lewis Publishers.

STOHS SJ, BAGCHI D, HASSOUN E, et al, 2000. Oxidative mechanisms in the toxicity of chromium and cadmium ions [J]. J. Environ. Pathol. Oncol, 19: 201-213.

TAMÁS L, VALENTOVICOVÁ K, HALUSKOVÁ L, et al, 2009. Effect of cadmium on the distribution of hydroxyl radical, superoxide and hydrogen peroxide in barley root tip [J]. Protoplasma, 236: 67-72.

VIARENGO A, CANESI L, PERTICA M, et al, 1990. Heavy metal effect on lipid peroxidation in the tissues of Mytilus galloprovincialis. L [J]. Comp. Biochem. Physiol (C), 97: 37-42.

VIEIRA LR, GRAVATO C, SOARES A, et al, 2009. Acute effects of copper and mercury on the estuarine fish Pomatoschistus microps: Linking biomarkers to behavior [J]. Chemosphere, 76: 1416-1427.

WANG N, INGERSOLLl CG, HARDESTY DK, et al, 2007a. Acute toxicity f copper, ammonia, and chlorine to glochidia and juveniles of freshwater mussels (Unionidae) [J]. Environ. Toxicol. Chem, 26: 2036-2047.

WANG N, INGERSOLL CG, GREER IE, et al, 2007b. Chronic toxicity testing of copper and ammonia to juvenile freshwater mussels (Unionidae) [J]. Environ. Toxicol. Chem, 26: 2048-2056.

第三章
饲料硒解除铜对皱纹盘鲍毒性作用的研究

第一节 引 言

硒是动植物必需的微量元素，是天然的重金属解毒剂（Frost，1979）。硒是体内抗氧化体系的组成部分，常作为含硒抗氧化酶（如谷胱甘肽过氧化物酶 GPx 和硫氧还蛋白还原酶 TPx 等）的活性中心而参与抗氧化作用（Wang 等，2007）。这些含硒的抗氧化酶连同抗氧化系统的其他成员（SOD、CAT、GPx、GST、GSH、TrxP 等）可以保护细胞免受由重金属诱导的自由基和过氧化脂质损伤（Finkel 和 Holbrook，2000），从而保护蛋白质和 DNA 及生物膜的完整性。贺宝芝等（1998）发现，硒可以显著提高铅中毒大鼠红细胞中的 SOD 酶活力，抑制红细胞脂质过氧化水平，并且对铅在骨、肾脏和肝脏中的积累也具有一定的降低作用。Trevisan 等（2011）研究表明，在 Cu 处理之前于水中预加一定量的硒，从很大程度上可以抵御紫贻贝（*Mytilus edulis*）蛋白巯基的过氧化和 DNA 的损伤。

硒可以解重金属毒性的另一条途径为硒能诱导金属硫蛋白（MT）的合成。郭军华等（1993）研究表明，硒能诱导 MT，且比之前认为对 MT 诱导能力最强的 Cd 的作用还好，后来 ELISA 法也证实了硒对 MT 确实有诱导作用，而且诱导效果高出 Zn 8～9 倍。李盟军（1995）等曾用实验证明了甘草锌（甘草的有效成分与锌的化合物)、富硒麦芽、亚硒酸钠能诱导金属硫蛋白，对顺铂所致小鼠生殖毒性有一定的保护作用。动物注射 Cd 或 Zn 盐能导致 Cd-MT 和 Zn-MT 在组织中的积累，这是由于 MT 的基因受离子诱导激活后，可以转录出大量的 MT-mRNA，从而指导合成多量 MT（周杰昊等，1995）。所以 MT 对过量的 Zn 和 Cu 有解毒作用，当 Zn 和 Cu 水平过量时，离子就以 MT 结合态的形式存在，这样有利于从组织中将这些金属清除出来，起到解除 Zn 和 Cu 毒害的作用。

大量的研究表明，水中的过量的 Cu 离子毒性机制之一为导致生物体的氧化应激，从而产生自由基，对生物体造成损伤。通过前面的实验也证实了这一毒性机制。在重金属污染越来越严重的今天，通过营养途径来降低重金属的毒性成为近年来的研究热点。本实验通过研究饲料中的营养素硒对皱纹盘鲍抗 Cu 的作用及机制，为重金属解毒提供可靠的数据支持。

第二节 摄食生长实验设计

（一）实验动物和养殖管理

皱纹盘鲍购自青岛鳌山卫育苗场，为人工孵化的同一批鲍苗。在中国海洋大学水产馆养殖系统中暂养 2 周后，挑选大小一致的健康个体［平均体重为 (3.17±0.01) g］随机分成 3 组，每组 3 个重复，每个重复 50 只鲍。生长实验在中国海洋大学水产馆养殖系统中进行，静水养殖 2 个月。实验期间，每天换水两次，每次换水量为实验缸水量的一半。每天 17:00 投喂人工饲料，次日 8:00 清底，并密切观察采食及健康状况。养殖过程中水温 18～21℃，盐度为 22～28，pH 为 7.4～7.9，溶氧量大于 6mg/L，水体中硒含量为 $0.46\mu g/L$。

（二）实验设计

根据第二章的实验中，只有 0.02mg/L Cu 和对照组的皱纹盘鲍在 28d 内没有死亡，故选择 0.02mg/L 作为本实验 Cu 的浓度。Cu 的实测值为 (0.018±0.01) mg/L。饲料中添加硒的水平为 0、1.5、4.5mg/kg。硒源为 $Na_2SeO_3 \cdot 5H_2O$。各饲料组中硒的实测值分别为 0.1、0.95、4.2mg/kg。基础饲料配方参照 Mai 等（2003），为半精制饲料，蛋白质源为明胶和酪蛋白，脂肪源为鲱鱼油和大豆油，两者的比例为 1∶1，主要糖源为糊精，再辅以纤维素、维生素和矿物质（不含硒）等配制而成。基础饲料中的常规指标（包括粗蛋白质、粗脂肪和粗灰分）的测定参照美国官方分析化学师协会（AOAC，1995）的方法。皱纹盘鲍的饲料配方及其营养成分见表 3-1。皱纹盘鲍饲料的具体配制步骤和保存方法参考 Zhang 等（2007）。

表 3-1 基础饲料配方及其营养成分

成分	含量（%）
酪蛋白[a]	25.00
明胶[b]	6.00
糊精[b]	33.50
羧甲基纤维素[b]	5.00
海藻酸钠[b]	20.00
维生素混合物[c]	2.00
矿物质混合物[d]	4.50
氯化胆碱[b]	0.50
豆油∶鲱鱼油[e]	3.50
概略养分分析（以干重计）	
粗蛋白质	29.41
粗脂肪	3.26
粗灰分	10.01

注：a. Sigma 公司。

b. 国药集团上海化学试剂有限公司。

c. 每 1 000g 饲料中含有：盐酸硫胺素，120mg；核黄素，100mg；叶酸，30mg；盐酸吡哆素，40mg；烟酸，800mg；泛酸钙，200mg；肌醇，4 000mg；生物素，12mg；维生素 B_{12}，0.18mg；维生素 C，4 000mg；维生素 E，450mg；维生素 K_3，80mg；维生素 A，100 000IU；维生素 D，2 000IU。

d. 每 1 000g 饲料中含有：NaCl，0.4g；$MgSO_4 \cdot 7H_2O$，6.0g；$NaH_2PO_4 \cdot 2H_2O$，10.0g；KH_2PO_4，20.0g；$Ca(H_2PO_4)_2 \cdot H_2O$，8.0g；Fe-柠檬酸，1.0g；$ZnSO_4 \cdot 7H_2O$，141.2mg；$MnSO_4 \cdot H_2O$，64.8mg；$CuSO_4 \cdot 5H_2O$，12.4mg；$CoCl_2 \cdot 6H_2O$，0.4mg；KIO_3，1.2mg。

e. 豆油∶鲱鱼油=1∶1。

（三）样品处理

养殖实验结束时，皱纹盘鲍禁食 3d 排空肠道内容物。实验缸中所有的皱纹盘鲍称重计数后，收集肝胰脏、肌肉、鳃、外套膜、贝壳。肝胰脏、肌肉、鳃和外套膜剪成小块混匀后分装在离心管中，保存在 −80℃ 冰箱中待测。离心得到的血清立即放在液氮中速冻后放 −80℃ 冰箱中保存。肝胰脏立即放入无 RNA 酶管中，液氮速冻后放入 −80℃ 冰箱保存。

肝胰脏样品在使用前解冻，加入预冷的 0.86% 生理盐水，冰上匀浆后 4℃、4 000r/min 离心 15min，取肝胰脏上清测定抗氧化指标。

以特定增长率（SGR）衡量皱纹盘鲍的生长情况：

$$SGR = 100 \times (\ln W_t - \ln W_i)/t$$

式中，W_t、W_i 分别代表鲍终末体重和初始体重（g），t 代表时间（d）。

第三节　饲料硒对铜胁迫下皱纹盘鲍生长和组织铜含量的影响

（一）对生长的影响

硒 1.5mg/kg 饲料组 [实测（1.30±0.13）mg/kg] 中硒经过 3h 和 6h 的溶失后硒含量分别为（1.04±0.12）、（0.95±0.05）mg/kg，均没有显著性差异。硒 4.5mg/kg 饲料组 [实测（4.17±0.15）mg/kg] 中硒经过 3、6h 的溶失后，其硒含量分别为（3.99±0.29）、（3.88±0.12）mg/kg，均没有显著性差异。如表 3-2 所示，饲料中不同的硒添加水平对 Cu 胁迫下皱纹盘鲍生长和存活率无显著性影响。各组皱纹盘鲍的存活率在 88.33%~98.33%；各组生长虽然没有统计学的显著性差异，但是随着硒含量的增加，SGR 呈增大的趋势。

（二）对组织 Cu 含量的影响

如表 3-3 所示，饲料中的 Se 显著影响皱纹盘鲍肝胰脏和血清中的 Cu 含量。1.5mg/kg 实验组肝胰脏中 Cu 含量显著低于对照组和 4.5mg/kg 组。添加 Se 实验组血清中 Cu 含量显著低于对照组。

鳃、外套膜、肌肉、贝壳中 Cu 的含量在各实验组虽然无显著性差异,但是 1.5mg/kg 组显著低于对照组和 4.5mg/kg 组。

表 3-2 饲料中的硒（Se）对 Cu 胁迫下的皱纹盘鲍生长和存活的影响（平均值±标准误，$n=3$）

饲料中硒含量 (mg/kg)	初始体重 (g)	终末体重 (g)	特定增长率 (%)	存活率 (%)
对照组	3.15±0.00	4.25±0.22	0.50±0.09	98.33±0.83
1.5	3.17±0.01	4.58±0.07	0.62±0.03	96.67±1.67
4.5	3.17±0.02	4.72±0.20	0.66±0.08	88.33±4.64
One-way ANOVA				
P 值	0.592	0.232	0.269	0.101
F 值	0.572	1.883	1.648	3.444

由于肝脏高的 Cu 积累量可导致肝脏细胞坏死是众所周知的 Cu 对动物的危害之一。本实验的研究发现，1.5mg/kg 实验组的肝胰脏的 Cu 含量显著低于对照组，且 1.5mg/kg 实验组在皱纹盘鲍的贝壳、肌肉、外套膜、鳃和血清中的 Cu 含量是所有的处理组中最低的。根据 Trevisan 等（2011）研究发现，在 Cu 离子胁迫下，预先用水中的硒处理（4μg/L 处理 3d）过的紫贻贝（*Mytilus edulis*）鳃组织中含有较低的 Cu 含量。根据 Richards 等（1989）和 Harris（1991）的研究发现，金属硫蛋白（MT）作为主要的细胞内的 Cu 结合蛋白，在降低组织 Cu 含量的过程中发挥重要的作用，因为 MT 结构中含有大量的半胱氨酸残基，这样使 Cu 和 MT 具有较高的结合能力（7～10g/mol），而且 MT 还是细胞内重要的抗氧化剂（Thornalley 和 Vasak，1985）。然而，MT 和 Cu 的结合能力还有其抗氧化力的能力都直接依赖于肝脏细胞中 GSH 的可利用率（Jiménez 和 Speisky，2000；Jiménez 等，2002）。而且有研究报道 GSH 的半胱氨基酸部分的巯基对 Cu、Hg、镉、铅具有较高的亲和力，形成具有较高稳定性的硫醇盐复合物，便于机体排出体外（Wang 和 Ballatori，1998）。本实验研究发现，添加 Se 的实验组的肝胰脏 GSH 的含量显著高于对照组，且 1.5mg/kg 实验组的 MT 含量显著高于对照组和 4.5mg/kg 实验组，推测 GSH 和 MT 在降低组织含量中发挥重要作用。

第三章 饲料硒解除铜对皱纹盘鲍毒性作用的研究

表3-3 饲料中的硒（Se）对Cu胁迫下的皱纹盘鲍组织Cu含量的影响（平均值±标准误，$n=3$）

饲料中硒含量 (mg/kg)	血清 (μg/mL)	贝壳 (μg/g)	肌肉 (μg/g)	外套膜 (μg/g)	鳃 (μg/g)	肝胰脏 (μg/g)
对照组	5.29±0.32[a]	19.49±3.69	8.76±0.37	15.28±1.88	13.92±1.27	13.66±0.49[a]
1.5	2.74±0.33[b]	16.39±1.92	7.98±0.45	14.81±1.84	13.96±2.44	9.08±0.51[b]
4.5	3.52±0.08[b]	21.14±3.35	9.18±0.48	16.93±1.52	16.45±1.06	13.44±0.58[a]
One-way ANOVA						
P 值	0.001	0.573	0.219	0.685	0.523	0.001
F 值	23.477	0.611	1.975	0.403	0.724	23.579

注：同列数据栏中，经Turkey检验差异不显著的平均值之间用相同的字母表示（$P>0.05$）。

然而在高硒组（4.5mg/kg），实验组在鳃、外套膜、贝壳和肌肉中 Cu 的含量要高于对照组。一些研究表明，硒元素并不能降低组织中的 Cd 含量（Meyer 等，1982；Stajn 等，1997；Jihen 等，2008）。Cerklewski（1976）研究发现，低剂量的硒可以降低小鼠组织 Pb 的含量，但是高剂量的硒使 Pb 的含量增加，研究结果和本实验中相似。实验得出含硒 1.5mg/kg 饲料组对改善组织 Cu 含量具有较好的作用，且优于含硒 4.5mg/kg 饲料组，具体的原因需要进一步的研究。

第四节　饲料硒对铜胁迫下皱纹盘鲍肝胰脏抗氧化指标的影响

（一）对抗氧化的影响

如表 3-4 所示，各处理组肝胰脏 SOD、CAT、GST、TrxP 的活力和 Trx 的含量无显著性差异。但是 1.5mg/kg Se 实验组肝胰脏 SOD、CAT 和 TrxP 的活力高于对照组和 4.5mg/kg 实验组。饲料中的 Se 显著影响了肝胰脏 Se-GPx、TrxR 的活力及 GSH 的含量。4.5mg/kg 饲料组 Se-GPx 显著低于对照组和 1.5mg/kg 组，但是 1.5mg/kg 饲料组与对照组无显著性差异。添加 Se 的饲料组的 TrxR 的活力和 GSH 的含量显著高于对照组。

重金属的毒性作用主要表现在产生对细胞有危害的过量的活性氧和降低细胞的抗氧化能力（Winterbourn，1982）。前期的研究发现，0.02mg/L 的 Cu 显著抑制了皱纹盘鲍肝胰脏抗氧化酶（SOD、GPx 和 GST）的活力（Lei 等，2015）。硒是人类和动物的必需微量元素，是一种抗氧化剂。硒的抗氧化作用主要表现在 Se 是一些抗氧化酶（如 Se-GPx 和 TrxR 等）的活性位点。许多研究发现，硒可以通过调节 GSH 的含量和抗氧化酶的活力来降低脂质过氧化的毒性（Combs 和 Combs，1984；McPherson，1994；Ji 等，2006）。Newairy 等（2007）和 Tran 等（2007）研究发现，硒的添加增强了机体抗氧化作用，从而降低了 Cu 或 Cd 的毒性。本研究

表 3-4 饲料中的硒（Se）对 Cu 胁迫下的皱纹盘鲍肝胰脏抗氧化相关指标的影响（平均值±标准误，$n=3$）

饲料中硒含量 (mg/kg)	SOD	CAT	Se-GPx	GST	GSH	Trx	TrxR	TrxP
对照组	10.77±0.38	6.42±1.53	16.58±1.02[a]	2.59±0.33	4.15±0.06[b]	289.83±0.38	2.20±0.06[b]	4.86±0.38
1.5	11.04±0.81	6.78±1.80	16.39±1.09[a]	2.31±0.15	5.13±0.21[a]	288.77±0.95	2.58±0.05[a]	5.10±0.39
4.5	10.79±0.44	5.63±1.61	10.51±1.67[b]	2.06±0.10	5.52±0.07[a]	286.80±0.66	2.36±0.02[b]	5.52±0.23
One-way ANOVA								
P 值	0.936	0.883	0.026	0.290	0.001	0.057	0.004	0.434
F 值	0.067	0.127	7.101	1.530	27.316	4.809	16.163	0.961

注：SOD，超氧化物歧化酶（U，每毫克蛋白中）；CAT，过氧化氢酶（U，每毫克蛋白中）；GPx，谷胱甘肽过氧化物酶（U，每毫克蛋白中）；GST，谷胱甘肽硫转移酶（U，每毫克蛋白中）；GSH，还原性谷胱甘肽（mg，每毫克蛋白中）；Trx，硫氧还蛋白（ng/mL）；TrxR，硫氧还蛋白还原酶（mU/L）；TrxP，硫氧还蛋白过氧化物酶（U，每毫克蛋白中）。同列数据栏中，经 Turkey 检验差异不显著的平均值之间用相同的字母表示（$P>0.05$）。

发现，1.5mg/kg 硒的实验组肝胰脏 SOD、CAT、TrxR、TrxP 的活力和 GSH 的含量要高于对照组，这表明这些抗氧化酶和抗氧化小分子 GSH 在饲料硒缓解 Cu 对皱纹盘鲍的毒性作用中发挥重要作用。

研究表明硒的处理可以升高哺乳动物、鱼类和水生无脊椎动物的含硒蛋白 Se-GPx 或 TrxR 的活力（Zhang 等，2008；Adesiyan 等，2011；Atencio 等，2009；Tran 等，2007；Wang 等，2010）。Tran 等（2007）研究发现，在 Cu 处理前给予硒预处理组的 Se-GPx 和 TrxR 含量显著升高。本实验研究发现，1.5mg/kg 硒实验组显著升高了 TrxR 的活性，但是对 Se-GPx 无显著性影响。众所周知，硒的添加能够升高硒蛋白酶活力的根本原因在于增加硒蛋白中硒半胱氨酸含量（Berggren 等，1999；Saito 和 Takahashi，2002）。虽然这两种抗氧化酶在催化中心都具有硒半胱氨酸残基，但是 Se-GPx 的硒半胱氨酸残基主要是谷氨酰胺和色氨酸，而 TrxR 的硒半胱氨酸残基是开放 C 端的三个相连的半胱氨酸残基。所以 TrxR 的活性位点的活泼性要比 Se-GPx 高。一些研究表明，TrxR 在皱纹盘鲍抵抗氧化应激中发挥重要作用（De Zoysa 等，2009）。实验表明，TrxR 在饲料硒解 Cu 毒性中发挥比 Se-GPx 更重要的作用。在实验中，4.5mg/kg 硒实验组显著降低 Se-GPx 的含量，说明高剂量的硒对抗氧化酶产生抑制的作用，相类似的结果在鱼类中也有发现（Tallandini 等，1996）。

（二）对氧化指标的影响

如表 3-5 所示，1.5mg/kg 组肝胰脏 MDA 的含量显著低于对照组，但是 4.5mg/kg 组 MDA 的含量和对照组无显著性差异。1.5mg/kg 组肝胰脏蛋白羰基含量显著低于对照组和 4.5mg/kg 组。各处理组肝胰脏 DNA 损伤（F 值）无显著性差异。

在氧化胁迫下产生的过量的 ROS 可引起脂质、蛋白质和 DNA 的氧化（Valentine 和 Wang，1998）。根据 Yuan 和 Tang（1999）研究报道，硒是人体的必需元素之一，它能够抵抗自由基，保护脂质、蛋白质或 DNA 结构的完整性。机体脂质过氧化的升高是导致氧化胁迫下细胞损伤的主要原因（Hermes-Lima 等，1995）。MDA 是细胞的脂质过氧化的主要产物。Chelomin 和 Belcheva（1991）研究表明，Cu 胁迫下扇贝（*Mizuhopecten yessoensis*）肝胰脏细胞中，随着 Cu 积累量的升高，MDA 的含量显著性升高。在本实验中，1.5mg/kg 硒处理组的 MDA 的含量显著低于对照组和 4.5mg/kg 处理组。Orun 等（2005）研究发现，硒的添加可以显著降虹鳟肝脏中的 MDA 含量。

表 3-5 饲料中的硒（Se）对 Cu 胁迫下皱纹盘鲍肝胰脏氧化指标的影响（平均值±标准误，$n=3$）

饲料中硒含量 (mg/kg)	MDA (nmol，每毫克蛋白中)	蛋白质羰基化 (nmol，每毫克蛋白中)	DNA 断裂程度 (F 值)
对照组	29.94±1.05[a]	8.67±0.23[a]	0.38±0.08
1.5	21.79±0.95[b]	4.35±0.32[b]	0.26±0.02
4.5	24.41±0.91[ab]	8.49±1.20[a]	0.25±0.01
One-way ANOVA			
P 值	0.016	0.009	0.186
F 值	8.918	11.232	2.254

ROS 导致很多非酶蛋白的改变，其中蛋白羰基化广泛被用作氧化应激的标志性指标（Shacter 等，1994；Je 等，2008）。羰基衍生物的生成是一个不可逆的过程，大大增加了蛋白被蛋白酶氧化的可能。Kaloyianni 等（2009）报道，在不同浓度金属和无机污染物的胁迫下，紫贻贝血细胞蛋白羰基显著升高。本实验研究发现，1.5mg/kg 硒实验组显著降低了肝胰脏蛋白羰基含量。Cu 的胁迫也可以导致 DNA 的损伤（Gabbianelli 等，2003）。Tran 等（2007）研究发现，硒的添加（4μg/L）可以显著降低紫贻贝（*Mytilus edulis*）Hg 导致的 DNA 的损伤。Trevisan 等（2011）也发现类似的结果，预处理硒可以显著降低 Cu 导致的蛋白氧化和

DNA 的损伤。然而，在本研究中，硒的添加对 DNA 的损伤无显著性影响。总之，硒的添加可以抵抗 Cu 胁迫下导致的脂质过氧化和蛋白羰基化，而且硒可以通过提高抗氧化酶的活力和抗氧化小分子的含量来对抗 Cu 导致的氧化损伤，尤其是对 TrxR 和 GSH 发挥重要作用。

第五节 饲料硒对铜胁迫下皱纹盘鲍肝胰脏金属硫蛋白（MT）含量和 MTF-1 mRNA 表达的影响

（一）金属硫蛋白（MT）和金属转录因子（MTF-1）的全长序列和结构分析

1. MT 基因全长 cDNA 序列及推测编码的氨基酸序列分析

皱纹盘鲍 MT 基因全长 cDNA 序列及推测编码的氨基酸序列见图 3-1。

```
ACATGGGATCGATATATAGCCCGAGTCAGCCTGACCTGCTTCAGTATCGCTGTTGAAGCTGTGTTTTGAAGAACAGAAGCGT
TTCTACCAGCCTCACTACATTATCATC
ATG TCC AGT CCC CAA GGC GCA GGA TGC ACA GGT GAG TGC AAG ACC GAC CCA TGC GCG TGC GGT
 M   S   S   P   Q   G   A   G   C   T   G   E   C   K   T   D   P   C   A   C   G
ACC GAC TGC AAG TGT AAC CCG GAC GAT TGT GCG TGC GAC TGC TGC AAG GTC AAG AAG TGC AAG
 T   D   C   K   C   N   P   D   D   C   A   C   D   C   C   K   V   K   K   C   K
TGT CCA GGT TCT TGC GAG TGT GGC AAA GGA TGC ACC AGT GGC GAG ACA TGC AAG TGT GAT GAC TCC
 C   P   G   S   C   E   C   G   K   G   C   T   S   G   E   T   C   K   C   D   D   S
TGC ACG TGC AAA
 C   T   C   K
TAGCCACGCCTCAAATAAATGCGCTCACATGACTTCCGATGACAGGAAGACACAGCATGAGAGGGACCAACTCTATTTTTTGT
CGCACTGTTACCTGCCGCTGGCATGAGATGGATGTGGAGTGATCTGGTGAAACAAACGTTCCAAGGTTGACAATTTGCAGAGA
CGCGCGGGAGGAGAACATTTTCACACAGCATATTCAAATCCCTTCAAGCACATTAAAGCTTTACATTGTTAAAAAAAAAAAAA
AAAAAAAAAAAAAA
```

图 3-1 皱纹盘鲍 MT 基因的核算序列全长及其推导的氨基酸序列

MT 基因全长 cDNA 序列包含 551 个核苷酸，108bp 的 5′非编码区、255bp 的 3′非编码区和 188bp 的开放阅读框。其中开放阅读框编码 69 个氨基酸，预测相对分子质量约为 6 996，理论等电点为 5.16。经 NCBI 蛋白 Blastp 比对发现，翻译之后的皱纹盘鲍 MT 的蛋白序列与九孔鲍（*Haliotis diversicolor supertexta*）、翡翠贻贝（*Perna viridis*）、四角蛤蜊（*Mactra quadrangularis*）、菲律宾蛤仔（*Ruditapes philippinarum*）的 MT 蛋白序列的同源性较高（同源性分别为 72%、67%、63%、60%）。利用 Bioedit 软件对 MT 的氨基酸组成进行分析发现，MT 分子富含 Cys（26.47%）、Ser（18.33%）、Lys（11.67%），不含组氨酸和芳香族氨基酸。皱纹盘鲍 MT 含 18 个 Cys 残基，而且它们与邻近氨基酸组成了金属硫蛋白保守性结构。MT 富含金属硫蛋白典型的 Cys-X（1~3）-Cys 结构，其中含有 7 个 Cys-X-Cys 结构，即 3 个 CKC、2 个 CAC、1 个 CEC、1 个 CTC；1 个 Cys-X（2）-Cys 结构，即 CDTC；5 个 Cys-X（3）-Cys 结构，即 CTGEC、CGTDC、CPGSC、CGKGC、CDDSC，并且皱纹盘鲍存在无脊椎动物 MT 的特征序列（CKCXXXCXCX）。使用 Bioedit 软件将皱纹盘鲍 MT 和其他物种进行多序列比对（图 3-2），可以看出皱纹盘鲍 MT 与其他物种的 MT 在 Cys 的排列方式上显示出较高的保守性。将皱纹盘鲍 MT 的氨基酸序列与来自其他 26 个物种的 MT 氨基酸序列用 MEGA4.0 软件进行 ClustalW 比对，构建了 MT 的 NJ 系统树（图 3-3）。在动物 MT 的进化簇中，爬行类、两栖类和鱼类聚为一支，鸟类为一支，软体动物和哺乳类聚为一支，皱纹盘鲍和魁蚶亲缘关系较近。

2. MTF-1 基因全长 cDNA 序列及推测编码的氨基酸序列分析

皱纹盘鲍 MTF-1 基因全长 cDNA 序列及推测编码的氨基酸序列见图 3-4。MTF-1 基因 cDNA 全长为 2 278bp，包括 25bp 的 5′非编码区、626bp 的 3′非编码区和 1 566bp 的开放阅读框，开放阅读框编码 522 个氨基酸，预测相对分子质量约为 59 630，理论等电点为 5.61。分析发现，氨基酸的结构域主要包括 6 个紧密相连的锌指结构域、1 个酸性结构域、1 个富含脯氨酸的结构域、1 个富含丝氨酸-苏氨酸结构域。

皱纹盘鲍的 MTF-1 的氨基酸序列与人（*Homo sapiens*）、斑马鱼（*Danio rerio*）、小鼠（*Mus musculus*）、河鲀（*Takifugu rubripes*）、果蝇（*Drosophila melanogaster*）的氨基酸序列进行

多序列比对，发现上述 5 个物种与皱纹盘鲍 MTF-1 的氨基酸序列一致性为 55%～70%，5 个锌指结构域保守性较高，序列相似性度达到 90%以上，而其他结构域相似性较低（图 3-5）。说明它们可能具有相似的结构和功能。

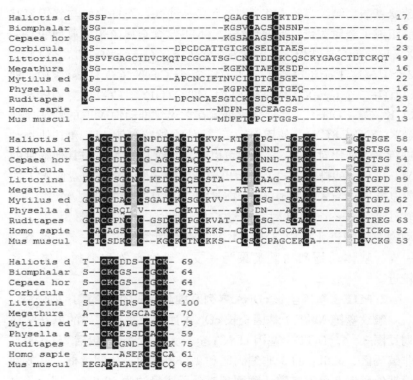

图 3-2 皱纹盘鲍 MT 的氨基酸多序列比对

注：相同的氨基酸用黑色背景显示。

研究人员对皱纹盘鲍和来自不同物种的 MTF-1 基因构建了系统进化树（图 3-6），并进行分子系统进化学的分析。结果显示，人和小鼠的聚在一支，河蚬和斑马鱼的聚在一支，皱纹盘鲍和它们的亲缘关系都不近。

第三章 饲料硒解除铜对皱纹盘鲍毒性作用的研究

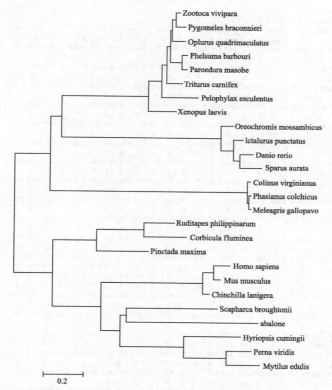

图 3-3 皱纹盘鲍（*Haliotis discus hannai* Ino.）
MT 蛋白与其他物种的进化树分析

Homo sapiens (AAP97267), *Chinchilla lanigera* (AAS59423), *Mus musculus* (AAH31758), *Danio rerio* (AAS00514), *Sparus aurata* (AAC32738), *Ictalurus punctatus* (AAC36348), *Oreochromis mossambicus* (AAB32778), *Phasianus colchicus* (CAA44369), *Meleagris gallopavo* (CAA44372), *Colinus virginianus* (CAA44371), *Xenopus laevis* (NP_001081042), *Triturus carnifex* (CCG27894), *Pelophylax esculentus* (CCG27895), *Zootoca vivipara* (CAJ32319), *Pygomeles braconnieri* (CAJ32318), *Phelsuma barbouri* (CAJ32317), *Paroedura masobe* (NP_001081042), *Oplurus quadrimaculatus* (CAJ32315), *Ruditapes philippinarum* (AEB91531), *Hyriopsis cumingii* (ACZ06027), *Perna viridis* (AAD02054), *Pinctada maxima* (ACJ22893), *Corbicula fluminea* (ABM55725), *Mytilus edulis* (CAE11861), *Scapharca broughtonii* (ACH99846)。

```
AAGCAGTGGTATCAACGCAGAGTAC
ATG GGG AGG CAA GAG TTT ACG TCC GGA AAA TTC TGT TCG GCC CAT TCG GTT TTC GCT ATG GAA
 M   G   R   Q   E   F   T   S   G   K   F   C   S   A   H   S   V   F   A   M   E
TTT ACT GAT AGC AAC GAC AGC AAT AAA GGA ACC GAT TCT TTT GAA GAT TTA ATG ACG TTT GTA GAC
 F   T   D   S   N   D   S   N   K   G   T   D   S   F   E   D   L   M   T   F   V   D
GAA GAT GGA CCC ACT GAC AGT GCA GTG ACC CAG GTA TAC ATA GAA AGA CTA GGA TCT CAC TCA
 E   D   G   P   T   D   S   S   A   V   T   Q   V   Y   I   E   R   L   G   S   H   S
AAA GAT GAC CAA AAC CAA GTG GCC GAT TCG GGG TCA GAT ACC ATC CCT CCT GGT ATG GAG AAC TGT
 K   D   D   Q   N   Q   V   A   D   S   G   S   D   T   I   P   P   G   M   E   N   C
CAA AAT GAA AGT TCT CTT TCA ACC ATC GAA CAT GAA GGG TAC ATA GAT AAT ACA ATA TCT GAC GAC
 Q   N   E   S   S   L   S   T   I   E   H   E   G   Y   I   H   N   T   I   S   D   D
CAG ATA TTG ATG ACC TTA AAC CCT GGA AAC GAG AGA ATG CCT ATC AAC CCG TCA CAT GCA ACC ATT
 Q   I   L   M   T   L   N   P   G   N   E   R   M   P   I   N   P   S   H   A   T   I
ACA CTG GAG ACA CAA GAT CCA TAC ACT AAT GCC AAG GAG GTG AAG AGG TTT CAA TGT AAC TTC CAA
 T   L   E   T   Q   D   P   Y   T   N   A   K   E   V   K   R   F   Q   C   N   F   Q
GAC TGT TCA AGA ACT TAC AGT ACA CCT GGC AAC CTG AAA ACA CAT CTG AAG ACC CAC CGA GGA GAA
 D   C   S   R   T   Y   S   T   P   G   N   L   K   T   H   L   K   T   H   R   G   E
TAT ACC TTT GTA TGT GAC GAG CAC GGC TGT GGG AAG GAG TTC CTC ACC TCA TAC AGT CTC AAA ATA
 Y   T   F   V   C   D   Q   H   G   C   G   K   E   F   L   T   S   Y   S   L   K   I
CAT GTT CGA GTA CAT ACA AAG GAA AAA CCT TAT GAG TGC GAC ACC ACC GGC TGT GAG AAG TCT TTC
 H   V   R   V   H   T   K   E   K   P   Y   E   C   D   T   T   G   C   E   K   S   F
AAC ACA CTT TAC AGG TTA CGT GCT CAC AAA CGT CTC CAC TCC GGG AAT ACT TTC AAC TGT GAT GAG
 N   T   L   Y   R   L   R   A   H   K   R   L   H   S   G   N   T   F   N   C   D   E
AGT GGA TGC ACA AAG TAC TTT ACT TTG AGT GAT CTT CGT AAA CAC ATT CGT ACA CAC ACT GGA
 S   G   C   T   K   Y   F   T   T   L   S   D   L   R   K   H   I   R   T   H   T   G
GAG AAA CCT TAT GTG TGC AGT GAA ACC GGC TGT CAA AAA GCT TTT GCT GCA AGC CAT CTA AAA
 E   K   P   Y   V   C   S   E   T   G   C   Q   K   A   F   A   A   S   H   H   L   K
ACA CAT TCT CGA CAT ACC TCA GGT GAG AAG CCA TAC ACA TGT TCC CAG GAG GGC CAC AAG TCT
 T   H   S   R   T   H   S   G   E   K   P   Y   T   C   S   Q   E   G   C   H   K   S
TTC ACA ACC AAC TAT AGC CTT AAG TCT CAC AAG AAC AGA CAT GAC AAG GGC GGA GGA CAG TCT GAT
 F   T   T   N   Y   S   L   K   S   H   K   N   R   H   D   K   G   G   Q   S   D
CCG TCA GGG ACA CAC GAA GCA GCG GAG ACA CAT GAC AGT GGA TCC ATG GCT GAA CAG CTT
 P   S   G   T   H   E   A   A   E   T   H   D   S   G   G   S   M   T   A   E   Q   L
TTC AAC ACT ATC TAT GTG AAT CCC ACT AGT ACG GAC CAT GTC AGC CTG GAC GAG GCT CTG CAA
 F   N   T   I   Y   V   N   P   T   S   T   D   H   V   S   L   D   E   A   A   L   Q
CAG ACA GAT GTT GTG CCA GGT ATA CAA ACA GTT CCT GTT CAA GAA ATA CTG CAG CCA GTG ATT CCT
 Q   T   D   V   V   P   G   I   Q   T   V   P   V   Q   E   I   L   Q   P   V   I   P
GTA GTG GAT ACT GGT GCA TCA GGA GCT CCC ACG CCC TCA GAG GGT TGT GTC GCA CAT GTT
 V   V   D   T   G   A   S   G   A   P   T   P   S   E   G   S   G   C   V   Q   H   V
ATT CTA AAC CAG TCA GCT ATA CCC ACC CTC TCA GAT GCA GCC ACA GAC TTC CTT CTT CCC AGC AGT
 I   L   N   Q   S   A   I   P   T   L   S   D   A   A   T   D   F   L   L   P   S   S
CTC AGC ACC TCC ACA CAC ACA GGA ACA CTT CCT GTC AGC AAC CGC TTG CAA GGG GAA GTA ACC CAG
 L   S   N   S   T   H   T   G   T   L   P   V   S   N   R   L   Q   G   E   V   T   Q
CCC CAG AAC ATT GTC CCA GCC CCA ACT ACA GAT GCA GTG CAA CTC CAA ATT CAG ACA TCT GGA
 P   Q   N   I   V   P   A   P   T   T   D   A   V   Q   L   Q   I   Q   T   A   S   G
AGC ACT GTT CCA GTC AGT CAG ATC TTT GTG CCA GTG GTT TCT AAC ACG GAC AAG GGT CCA GTC ATA
 S   T   V   P   V   S   Q   I   F   V   P   V   V   S   N   T   D   K   G   P   V   I
GAG CTT GTG CCA CTT CAG AAC AGC ATT TCA GTG AAC GAT GAA AGC CGC ACA
 E   L   V   P   L   Q   N   S   I   S   V   N   D   E   S   R   T
TGAAATGCCACATAAACCCTATGCTTCAAGATTTTCACCTTTTTGTGGAATTTATGTGTTGATAATTTTCATGTTGTTTTGT
CTTCACACCCCAGGACAATGTGTGCATATTTCTATGTGTCAGCTTGTTGCAGATGGCTGCCTGCTTTGTGCACTGTGTTTAAT
ACCTACTATCTACTACACATTCATAACTTGCCAAATCTTTAGTTCAAATTGAGCCATTCACCATAAACTGCTTTTGTCTTCAG
GATGCATGGTTCCATGTGTCACATTACTGATGGCAGATAGGAGGTATGTGAATACATCTAAATGAAAACTTGCCTGAAAAGTA
AATGTTCTGGAATAATCTACTTACACATTTCATTTGCAATATTATCTTTATTCATTAGCAGCCTGTTTTTTGGGTGTAAGTT
CCATTACATACCCTAACCACCAGAACCAACACCCATCAACTTCCCCACCCCACATTACCAAACTCATTACAGTCAGATTTCA
CTTAAATATGGCAACTTTCTGTGTGAAAATCTAAGGACACCTCCATAAGGAGTCACTTCAGGGGAGCTTTTTAATCCGCCAAT
CAGCCTCAAGGCTTTTAAATTCTCAGTATGAGTAAGCAAATATTCAGACATGCATGTGTGATTTCTGTATATGGAAACAAA
TATAATCAGGGTCTTCCAACTGCCAAAAAAAAAAAAAAAAAAAAAAAAAAAA
```

图 3-4 皱纹盘鲍 MTF-1 基因的核酸序列全长及其推导的氨基酸序列

第三章 饲料硒解除铜对皱纹盘鲍毒性作用的研究

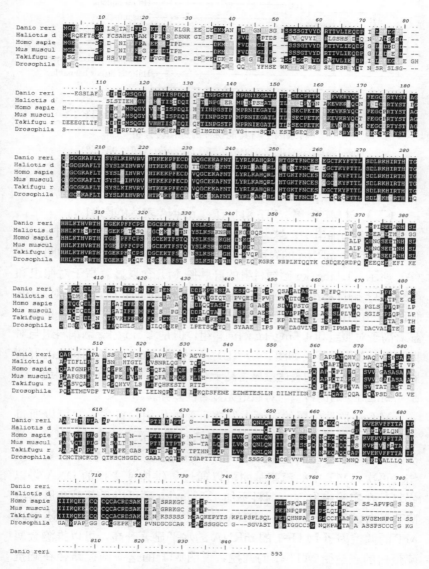

图 3-5 皱纹盘鲍 MTF-1 的氨基酸多序列比对
注：相同的氨基酸用黑色背景显示。

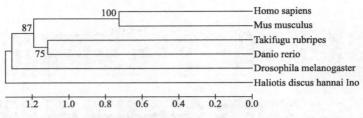

图 3-6 皱纹盘鲍（*Haliotis discus hannai* Ino.）MTF-1
转录因子与其他物种的进化树分析

Homo sapiens（NP_005946），*Mus musculus*（NP_032662），
Takifugu rubripes（NP_001027866），*Danio rerio*（NP_694513），
Drosophila melanogaster（ABW08505）。

（二）对 MT 含量、MT 和 MTF-1 mRNA 表达的影响

在硒各处理组中，笔者团队发现，皱纹盘鲍肝胰脏中 MT 含量、MT 和 MTF-1 mRNA 的表达水平呈现先升高后降低的趋势。如图 3-7（A）所示，1.5mg/kg 硒饲料组的肝胰脏 MT 含量显著高于对照组和 4.5mg/kg 硒饲料组，且 4.5mg/kg 硒饲料组的 MT 显著高于对照组。如图 3-7（B）所示，1.5mg/kg 硒饲料组的肝胰脏 MT mRNA 的表达显著高于对照组和 4.5mg/kg 硒饲料组，但是 4.5mg/kg 硒饲料组与对照组无显著性差异。如图 3-7（C）所示，1.5mg/kg 硒饲料组的肝胰脏 MTF-1 mRNA 的表达量显著高于对照组和 4.5mg/kg 硒饲料组，但是 4.5mg/kg 硒饲料组与对照组无显著性差异。

MT 在抗重金属的应激中发挥重要的作用。首先，它可以通过巯基和自由基进行结合，降低重金属过程中产生的过量 ROS。其次，MT 可以与多种重金属结合，减轻重金属对生物体的氧化损伤。再次，MT 中的巯基基团可以通过氢供体使受损的 DNA 得以修复。在实验中发现，饲料添加 1.5mg/kg 硒可以显著升高肝胰脏 MT 蛋白的含量及 MT mRNA 的表达。另外，笔者团队还发现 MT 的调控因子 MTF-1 的 mRNA 的水平显著升高，可以推测金属硫蛋白的表达受 MTF-1 的调控，说明 MT 的这一调控路径在硒抗 Cu 的氧化胁迫及降低重金属的含量中发挥重要的作用，但是具体的机制还需要进一步的研究。

第三章 饲料硒解除铜对皱纹盘鲍毒性作用的研究

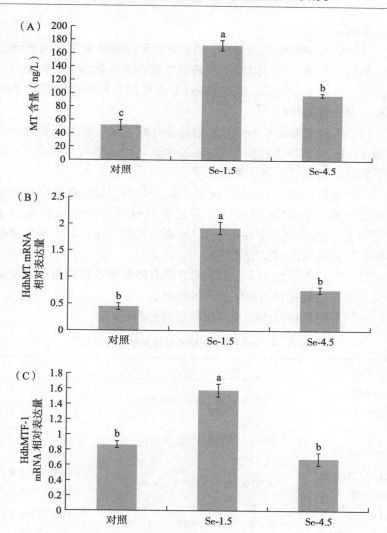

图 3-7 不同 Se 元素处理对皱纹盘鲍 Cu 胁迫下的肝胰腺中
MT 含量、HdhMT 和 HdhMTF-1 基因的表达的影响

(A) 不同浓度下的 Se 对 Cu 胁迫下皱纹盘鲍肝胰脏 MT 含量的影响；(B) 不同浓度下的 Se 对 Cu 胁迫下皱纹盘鲍肝胰脏 MT mRNA 表达的影响；(C) 不同浓度下的 Se 对 Cu 胁迫下皱纹盘鲍肝胰脏 MTF-1 mRNA 表达的影响。

小结：

(1) 饲料中添加 1.5mg/kg 的硒可以降低皱纹盘鲍肝胰脏和血清中的组织含量，并可以降低肝胰脏中脂质过氧化的水平和蛋白羰基含量，说明 1.5mg/kg 硒对减轻 Cu 对皱纹盘鲍的氧化损伤和毒性有一定的缓解作用。

(2) 饲料中添加 1.5mg/kg 硒显著升高了皱纹盘鲍肝胰脏抗氧化酶 TrxR、TrxP 的活性及 GSH 的含量，说明硒可以提高 Cu 胁迫下的皱纹盘鲍肝胰脏抗氧化水平。

(3) 饲料中添加 1.5mg/kg 的硒显著升高了皱纹盘鲍肝胰脏金属硫蛋白及其 mRNA 的表达，并显著诱导其上游金属转录因子 MTF-1 的表达。这表明在硒缓解 Cu 毒性的过程中，MT 发挥重要作用，具体机制还需要继续研究。

(4) 饲料中添加 4.5mg/kg 硒并没有降低皱纹盘鲍组织中金属含量及脂质过氧化和蛋白羰基化的水平。

本研究中所用的引物及其核酸序列见表 3-6。

表 3-6 本研究中所用的引物及其核酸序列

引物名	序列 (5′→3′)	序列信息
HdhMT (forward) 01	ATGTCCAGTCCCCAAGGC	HdhMT RT primer
HdhMT (reverse) 01	CCACACTCGCAAGAACCTG	HdhMT RT primer
HdhMTF-1 (forward) 01	GGSTGYACVCGSACCTACAG	HdhMTF-1 RT primer
HdhMTF-1 (reverse) 01	AYGGCTYTCWCCWGTRTGTGTCC	HdhMTF-1 RT primer
HdhMT (forward) 02	CCACACTCGCAAGAACCTGGACACT	3′ RACE primer
HdhMT (forward) 03	CTCCACACTCGCAAGAACCTGGACA	3′ RACE primer

第三章 饲料硒解除铜对皱纹盘鲍毒性作用的研究

（续）

引物名	序列（5′→3′）	序列信息
HdhMT (reverse) 04	CCACACTCGCAAGAACCTGGACACT	5′ RACE primer
HdhMT (reverse) 05	CGCATTTATTTGAGGCGTGGCTATT	5′ RACE primer
HdhMTF-1 (forward) 02	TTCAACACACTTTACAGGTTACGTGCTC	3′ RACE primer
HdhMTF-1 (forward) 03	CAAAAGAAAAACCTTATGAGTGCGACA	3′ RACE primer
HdhMTF-1 (reverse) 04	TTTGTGCATCCACTCTCATCACAGTTG	5′ RACE primer
HdhMTF-1 (reverse) 05	TGAAAGACTTCTCACAGCCGGTGGTGT	5′ RACE primer
UPM	AAGCAGTGGTATCAACGCAGAGTACGCGGGG	5′ RACE primer
NPU	AAGCAGTGGTATCAACGCAGAGT	5′ RACE primer
HdhMT (forward) 06	ATGTCCAGTCCCCAAGGC	MT Real-time RT primer
HdhMT (reverse) 06	CCACACTCGCAAGAACCTG	MT Real-time RT primer
HdhMTF-1 (forward) 06	CGGCTGTGAGAAGTCTTTCAAC	MTF-1 Real-time RT primer
HdhMTF-1 (reverse) 06	TGTCCGAATGTGTTTACGAAGATC	MTF-1 Real-time RT primer
β-actin (forward)	ACTCATTCACCACCACCG	β-actin Real-time primer
β-actin (reverse)	GGATGAAGAGGCAGCAGTAG	β-actin Real-time primer
M13 (forward)	GAGCGGATAACAATTTCACACAGG	Vector primer
M13 (reverse)	CGCCAGGGTTTTCCCAGTCACGAC	Vector primer

参 考 文 献

郭军华，徐卓立，吴德政，1993. 诱导生物合成金属硫蛋白减轻顺铂对小鼠的致死毒性 [J]. 癌症，12（6）：476.

贺宝芝，徐臻，1998. 有机硒对铅中毒的拮抗作用试验研究 [J]. 卫生研究，27（4）：229-232.

李盟军，郭军华，徐卓立，等，1995. 甘草锌、富硒麦芽和亚硒酸钠诱导金属硫蛋白对顺铂所致生殖毒性的影响 [J]. 军事医学科学院院刊（3）：239-240.

周杰昊，程时，1995. 金属硫蛋白与医学 [J]. 生理科学进展，26（1）：29-34.

ADESIYAN AC, OYEJOLA TO, ABARIKWU SO, et al, 2011. Selenium provides protection to the liver but not the reproductive organs in an atrazine-model of experimental toxicity [J]. Exp. Toxicol. Pathol, 63: 201-207.

ATENCIO L, MORENO I, JOS A, et al, 2009. Effects of dietary selenium on the oxidative stress and pathological changes in tilapia (*Oreochromis niloticus*) exposed to a microcystin-producing cyanobacterial water bloom [J]. Toxicon, 53: 269-282.

BERGGRE MM, MANGIN JF, GASDAKA JR, et al, 1999. Effect of selenium on rat thioredoxin reductase activity: increase by supranutritional selenium and decrease by selenium deficiency [J]. Biochem. Pharmacol, 57(2): 187-193.

CERKLEWSKI FL, FORBES RM, 1976. Influence of dietary zinc on lead toxicity in rats [J]. J. Nutr, 106 (5): 689-696.

CHELOMIN VP, BELCHEVA NN, 1991. Alterations of microsomal lipid synthesis in gill cells of bivalve mollusk *Mizuhopecten yessoensis* in response to cadmium accumulation [J]. Comp. Biochem. Physiol (C), 99: 1-5.

COMBS GF, COMBS SB, 1984. The nutritional biochemistry of selenium [J]. Ann. Rev. Nutr, 4: 257-80.

ZOYSA MD, NIKAPITIYA C, JEON Y J, et al, 2008. Anticoagulant activity of sulfated polysaccharide isolated from fermented brown seaweed Sargassum fulvellum [J]. J. Appl. Phycol, 20 (1): 67-74.

DE ZOYSA M, WHANG I, LEE Y, et al, 2009. Transcriptional analysis of antioxidant and immune defense genes in disk abalone (*Haliotis discus discus*) during thermal, low-salinity and hypoxic stress [J]. Comp. Biochem. Physiol.

(B), 154: 387-395.

FINKEL T, HOLBROOK NJ, 2000. Oxidants, oxidative stress and the biology of ageing [J]. Nature, 408: 239-247.

FROST DV, 1979. Selenium and vitamin E as antidotes to heavy metal toxicitiess [M] //SPALLHOLZ J E, Selenium in Biology and Medicine. Westport: AVC.

GABBIANELLI R, LUPIDI G, VILLARINI M, et al, 2003. DNA damage induced by copper on erythrocytes of gilthead sea bream *Sparus aurata* and mollusk *Scapharca inaequivalvis* [J]. Arch. Environ. Contam. Toxicol, 45: 350-356.

HARRIS ED, 1991. Copper transport: an overview [J]. Exp. Biol. Med, 196 (2): 130-140.

HERMES-LIMA M, WILLMORE WG, STOREY KB, 1995. Quantification of lipid peroxidation in tissue extracts based on Fe (III) xylenol orange complex formation [J]. Free Radical Bio. Med, 19: 271-280.

JE JH, LEE TH, KIM DH, et al, 2008. Mitochondrial ATP synthase is a target for TNBS-induced protein carbonylation in XS-106 dendritic cells [J]. Proteomics, 8: 2384-2393.

JI X, WANG W, CHENG J, et al, 2006. Free radicals and antioxidant status in rat liver after dietary exposure of environmental mercury [J]. Environ. Toxicol. Pharmacol, 22: 309-314.

JIHEN EH, IMED M, FATIMA H, et al, 2008. Protective effects of selenium (Se) and zinc (Zn) on cadmium (Cd) toxicity in the liver and kidney of the rat: histology and Cd accumulation [J]. Food Chem. Toxicol, 46: 3522-3527.

JIMÉNEZ I, ARACENA P, LETELIER ME, 2002. Chronic exposure of HepG2 cells to excess copper results in depletion of glutathione and induction of metallothionein [J]. Toxicology in vitro, 16 (2): 167-175.

JIMÉNEZ I, SPEISKY H, 2000. Effects of copper ions on the free radicalscavenging properties of reduced glutathione: implications of a complex formation [J]. Trace Elem. Biol. Med, 14: 161-167.

KALOYIANNI M, DAILIANIS S, CHRISIKOPOULOU E, et al, 2009. Oxidative effects of inorganic and organic contaminants on haemolymph of mussels [J]. Comp. Biochem. Physiol. Part C, 149: 631-639.

LEI Y, ZHANG W, XU W, et al, 2015. Effects of waterborne Cu and Cd on anti-oxidative response, lipid peroxidation and heavy metals accumulation in abalone Haliotis discus hannai ino [J]. Journal of Ocean University of China, 14 (3): 511-521.

MAI K, ZHANG W, TAN B, et al, 2003. Effects of dietary zinc on the shell biomineralization in abalone *Haliotis discus hannai* Ino [J]. J. Exp. Mar. Biol. Ecol, 283: 51-62.

MCPHERSON A, 1994. Selenium vitamin E and biological oxidation [M] // COLE DJ, GARNSWORTHY PJ. Recent advances in animal nutrition. Oxford: Butterworth and Heinemann's.

MEYER SA, HOUSE WA, WELCH RM, 1982. Some metabolic interrelationships between toxic levels of cadmium and nontoxic levels of selenium fed to rats [J]. J. Nutr, 112 (5): 954-961.

NEWAIRY AA, EL-SHARAKY AS, BADRELDEEN MM, et al, 2007. The hepatoprotective effects of selenium against cadmium toxicity in rats [J]. Toxicology, 242: 23-30.

ORUN I, ATES B, SELAMOGLU Z, et al, 2005. Effects of various sodium selenite concentrations on some biochemicaland hematological parameters of rainbow trout (*Oncorhynchus mykiss*) [J]. Fresen. Environ. Bull, 14: 18-22.

RICHARDS M, 1989. Recent developments in trace element metabolism and function: role of metallthionein in copper and zinc metabolism [J]. J. Nutr, 119: 1062-1070.

SAITO Y, TAKAHASHI K, 2002. Characterization of selenoprotein P as a selenium supply protein [J]. Eur. J. Biochem, 269: 5746-5751.

SHACTER E, WILLIAMS JA, LIM M, et al, 1994. Differential susceptibility of plasma proteins to oxidative modification: examination by western blot immunoassay [J]. Free Radical Bio. Med, 17: 429-437.

STAJN A, ZIKIĆ RV, OGNJANOVIĆ B, et al, 1997. Effect of cadmium and selenium on the antioxidant defense system in rat kidneys [J]. Comp. Biochem. Physiol. C Pharmacol. Toxicol. Endocrinol, 117 (2): 167-172.

TALLANDINI L, CECCHI R, DE BONI S, et al, 1996. Toxic levels of selenium in enzymes and selenium uptake in tissues of a marine fish [J]. Biol. Trace Elem. Res, 51: 97-106.

THORNALLEY PJ, VASAK M, 1985. Possible role for metallothionein in protection against radiation-induced oxidative stress. Kinetics and mechanism of its reaction with superoxide and hydroxyl radicals [J]. Biochim. Biophys. Acta, 827: 36-44.

TRAN D, MOODY AJ, FISHER AS, 2007. Protective effects of selenium on mercury-induced DNA damage in mussel haemocytes [J]. Aquat. Toxicol, 84: 11-18.

TREVISAN R, MELLO DF, FISHER AS, et al, 2011. Selenium in water enhances antioxidant defenses and protects against copper-induced DNA damage in the blue mussel *Mytilusedulis* [J]. Aquat. Toxicol, 101: 64-71.

VALENTINE R L, WANG H, 1998. Iron Oxide Surface Catalyzed Oxidation of Quinoline by Hydrogen Peroxide [J]. J. Environ. Eng, 124 (1): 31-38.

WANG HW, XU HM, XIAO GH, et al, 2010. Effects of selenium on the antioxidant enzymes response of Neocaridina heteropoda exposed to ambient nitrite [J]. Bull. Environ. Contam. Toxicol, 84: 112-117.

WANG W, BALLATORI N, 1998. Endogenous glutathione conjugates: occurrence and biological functions. Pharmacol Rev 50: 335-356 [J]. Pharmacol. Rev, 50 (3): 335-356.

WINTERBOURN CC, 1982. Superoxide-dependent production of hydroxyl radicals in the presence of iron salts (Letter) [J]. Biochem. J, 205: 463.

YUAN X, TANG C, 1999. Lead effect on DNA and albumin in chicken blood and the protection of selenium nutrition [J]. J. Environ. Sci. Health A, 34 (9): 1875-1887.

ZHANG JS, WANG HL, PENG DG, 2008. Further insight into the impact of sodium selenite on selenoenzymes: high-dose selenite enhances hepatic thioredoxin reductase 1 activity as a consequence of liver injury [J]. Toxicol. Lett, 176: 223-229.

ZHANG W, MAI K, XU W, 2007. Interaction between vitamins A and D on growth and metabolic responses of abalone *Haliotis discus hannai*, Ino [J]. J. Shellfish Res, 26: 51-58.

第四章
饲料硫辛酸解除铜对皱纹盘鲍毒性作用的研究

第一节 引 言

α-硫辛酸（alpha-lipoic acid，α-LA）是一种被列为 B 族维生素的抗氧化剂（Reed，2001）。LA 具有特殊的双硫五元环结构，与自由基反应的能力很强。LA 可以清除因 Cu 氧化胁迫而导致的过量自由基，如过氧化氢（H_2O_2）、羟自由基（·OH）、单线态氧（$^1O_2$）和次氯酸（HClO）等（Navari 等，2002；田芳等，2007）。LA 除了本身具有抗氧化作用之外，还可以与其代谢产物二氢硫辛酸（DHLA）联合一起清除生物体内过多的活性氧，并且可以激活生物体内其他抗氧化剂的代谢循环，它们相互补充，相互协调，共同发挥抗氧化作用（Packer 等，1995）。LA 和 DHLA 在机体内的相互转化可以还原细胞中一些抗氧化剂，例如维生素 C、维生素 E、GSH、硫氧还原蛋白和泛醌，使它们由氧化型转化为还原型。通过细胞培养发现，DHLA 能提高半胱氨酸的吸收，从而导致 GSH 的合成增加（Kramer 等，2001）。所以 LA 和 DHLA 被称为万能的抗氧化剂。

研究表明，LA 可以提高体内的抗氧化酶的水平（Arivazhagan 等，2001；Carey 等，2002）。在小鼠中的研究表明，外源补充 LA 可以显著性降低组织内脂质过氧化的水平，提高心肌和肝脏组织中的 GSH 含量，并提高 GPx、GST、GR 活性，提高机体抗氧化状态（李奕，2006）。由于其较好的抗氧化能力，LA 在抗应激和抗氧化胁迫下中的作用得到了广泛的关注，如顺铂造成的耳蜗氧化损伤（Leonard 等，1999）、缺血再灌注损伤（Cao 等，2003；刘学忠等，2004）、糖尿病所致的视网膜机能异常（Kowluru 等，2005）、阿尔茨海默病（Zhang 等，2001；Quinn 等，2007）等。在水产动物的研究中，Monserrat 等（2008）发现，LA 能够显著提高胡椒甲鲶（*Corydoras paleatus*）的抗氧化能力。

大量的研究表明，水中过量的 Cu 离子可以导致生物体的氧化应激，从而产生自由基，对生物体造成损伤。目前，关于 LA 本身强大的抗氧化能力对于重金属氧化胁迫下的抵抗作用方面的研究甚少。通过营养途径来降低重金属的毒性成为近年来的研究热点。本实验通过研究饲料中的抗氧化剂硫辛酸对皱纹盘鲍抗 Cu 的作用及机制，为重金属解毒提供可靠的数据支持。

第二节 摄食生长实验设计

（一）实验动物和养殖管理设计

皱纹盘鲍幼鲍为人工孵化同一批鲍，购自青岛鳌山卫育苗场。在中国海洋大学水产馆养殖系统中暂养 2 周后挑选大小一致的健康个体［平均体重为（3.17 ± 0.01）g］，随机分成 3 组，每组 3 个重复，每个重复 50 只鲍。生长实验在中国海洋大学水产馆养殖系统中进行，静水养殖 2 个月。实验期间，每天换水 2 次，每次换水量为实验缸水量的一半。每天 17：00 投喂人工饲料，次日 8：00 清底，并密切观察采食及健康状况。养殖过程中水温 18~21℃，盐度为 22~28，pH 为 7.4~7.9，溶氧量大于 6mg/L。

（二）养殖实验设计

根据第二章的实验，笔者团队发现只有 0.02mg/L Cu 和对照组的皱纹盘鲍在 28d 内没有死亡，故选择 0.02mg/L 的 Cu 作为本实验 Cu 的浓度。Cu 的实测值为（0.018 ± 0.01）mg/L。饲料中添加硫辛酸的水平为 0、700、2 100mg/kg，硒源为 $Na_2SeO_3 \cdot 5H_2O$。基础饲料配方参照 Mai 等（2003），为半精制饲料，蛋白质源为明胶和酪蛋白，脂肪源为鲱鱼油和大豆油，两者的比例为 1∶1，主要糖源为糊精，再辅以纤维素、维生素和矿物质（不含硒）等配制而成。基础饲料中的常规指标，包括粗蛋白质、粗脂肪和粗灰分的测定参照 AOAC（1995）。皱纹盘鲍的饲料配方及其营养成分见表 4-1。皱纹盘鲍饲料的具体配制步骤和保存方法参考 Zhang 等（2007）。

表 4-1 基础饲料配方及其营养成分

成分	含量（%）
酪蛋白[a]	25.00
明胶[b]	6.00
糊精[b]	33.50
羧甲基纤维素[b]	5.00
海藻酸钠[b]	20.00
维生素混合物[c]	2.00
矿物质混合物[d]	4.50
氯化胆碱[b]	0.50
豆油∶鲱鱼油[e]	3.50
概略养分分析（以干重计）	
粗蛋白质	30.81
粗脂肪	3.85
粗灰分	11.01

注：a. sigma 公司。

b. 国药集团上海化学试剂有限公司。

c. 每 1 000g 饲料中含有：盐酸硫胺素，120mg；核黄素，100mg；叶酸，30mg；盐酸吡哆素，40mg；烟酸，800mg；泛酸钙，200mg；肌醇，4 000mg；生物素，12mg；维生素 B_{12}，0.18mg；维生素 C，4 000mg；维生素 E，450mg；维生素 K_3，80mg；维生素 A，100 000IU；维生素 D，2 000IU。

d. 每 1 000g 饲料中含有：NaCl，0.4g；$MgSO_4 \cdot 7H_2O$，6.0g；$NaH_2PO_4 \cdot 2H_2O$，10.0g；KH_2PO_4，20.0g；$Ca(H_2PO_4)_2 \cdot H_2O$，8.0g；Fe-柠檬酸，1.0g；$ZnSO_4 \cdot 7H_2O$，141.2mg；$MnSO_4 \cdot H_2O$，64.8mg；$CuSO_4 \cdot 5H_2O$，12.4mg；$CoCl_2 \cdot 6H_2O$，0.4mg；KIO_3，1.2mg，$Na_2SeO_3 \cdot 5H_2O$，1.0mg。

e. 豆油∶鲱鱼油=1∶1。

(三) 取样与样品处理

养殖实验结束时，皱纹盘鲍禁食3d，排空其肠道的内容物。实验缸中所有皱纹盘鲍称重计数后，收集肝胰脏、肌肉、鳃、外套膜、贝壳。肝胰脏、肌肉、鳃和外套膜剪成小块混匀后分装在离心管中，保存在 $-80℃$ 冰箱中待测。离心得到的血清立即放在液氮中速冻后放 $-80℃$ 冰箱中保存。肝胰脏立即放入无RNA酶管中，液氮速冻后放入 $-80℃$ 冰箱中保存。

肝胰脏样品使用前解冻，加入预冷的0.86%生理盐水中，冰上匀浆，然后4℃，4 000r/min离心15min，取肝胰脏上清测定抗氧化指标。

以特定增长率（SGR）衡量鲍鱼的生长情况：

$$SGR = 100 \times (\ln W_t - \ln W_i) / t$$

式中，W_t、W_i 分别代表鲍终末体重和初始体重（g）；t 代表时间（d）。

第三节 饲料中的硫辛酸对铜胁迫下皱纹盘鲍生长存活和组织铜含量的影响

(一) 对生长存活的影响

如表4-2所示，饲料中添加不同的LA水平对Cu胁迫下皱纹盘鲍生长没有产生显著性影响。在Cu胁迫下，皱纹盘鲍的特定增长率（SGR）随着饲料中LA的增加而升高，但与对照组并无显著性差异。

表4-2 饲料中的硫辛酸（LA）对Cu胁迫下的皱纹盘鲍生长和存活的影响（平均值±标准误，$n=3$）

饲料中LA含量 (mg/kg)	初始体重 (g)	终末体重 (g)	特定增长率 (%)	存活率 (%)
对照	3.18±0.01	4.49±0.14	0.50±0.08	92.50±2.89
700	3.17±0.05	4.31±0.09	0.51±0.03	95.83±0.83

(续)

饲料中 LA 含量 (mg/kg)	初始体重 (g)	终末体重 (g)	特定增长率 (%)	存活率 (%)
2 100	3.15±0.01	4.46±0.08	0.58±0.02	98.33±0.83
One-way ANOVA				
P 值	0.376	0.514	0.507	0.150
F 值	1.155	0.745	0.604	2.643

注：同列数据栏中，经 Turkey 检验差异不显著的平均值之间用相同的字母表示（$P>0.05$）。

各组皱纹盘鲍的存活率在 92.50%～98.33%；各组生长虽然没有显著性差异，但是随着 LA 含量的升高，存活率呈增加的趋势。

（二）对组织金属含量的影响

如表 4-3 所示，饲料中的 LA 显著性影响皱纹盘鲍血清、肌肉、外套膜、鳃和肝胰脏中的 Cu 含量。700、2 100mg/kg 实验组血清、外套膜、鳃和肝胰脏中 Cu 含量显著低于对照组。2 100mg/kg 实验组肌肉中的 Cu 含量显著低于对照组，700mg/kg 实验组肌肉中的 Cu 含量虽然低于对照组，但是无显著性差异。

肝脏高的 Cu 积累量导致肝脏细胞坏死是众所周知的 Cu 对生物体的危害之一。本实验研究发现，700、2 100mg/kg 实验组降低了皱纹盘鲍血清、肝胰脏、肌肉、外套膜和鳃中的 Cu 含量。LA 具有螯合金属离子的能力，直接与 Cu^{2+} 螯合形成亲脂性复合物（Ou 等，1995），有利于 Cu 离子的排出和减轻 Cu 离子的危害。根据 Richards 等（1989）和 Harris（1991）的研究发现，金属硫蛋白（MT）作为主要的细胞内的 Cu 结合蛋白，在降低组织 Cu 含量的过程中发挥重要的作用，因为 MT 氨基酸结构中含有大量的半胱氨酸残基，从而使 Cu 和 MT 具有较高的结合能力（7～10g/mol）（Thornalley 和 Vasak，1985）。然而，MT 和 Cu 的结合能力还有其抗氧化力都直接依赖于肝脏细胞中 GSH 的可利用率

第四章 饲料硫辛酸解除铜对皱纹盘鲍毒性作用的研究

表4-3 饲料中的硫辛酸（LA）对 Cu 胁迫下的皱纹盘鲍组织 Cu 含量的影响（平均值±标准误，$n=3$）

饲料中 LA 含量 (mg/kg)	血清 (μg/mL)	贝壳 (μg/g)	肌肉 (μg/g)	外套膜 (μg/g)	鳃 (μg/g)	肝胰脏 (μg/g)
对照	10.69±1.41^a	5.74±0.36	13.61±0.19^a	22.58±0.55^a	18.47±0.35^a	27.05±1.65^a
700	4.99±0.49^b	6.91±0.40	10.07±1.17^{ab}	15.13±0.78^b	14.99±1.06^b	9.17±1.76^b
2 100	4.56±1.41^b	7.08±0.41	8.19±1.40^b	15.52±0.50^b	13.28±0.54^b	13.34±0.72^b
One-way ANOVA						
P 值	0.019	0.097	0.029	0.000	0.006	0.000
F 值	8.327	3.528	6.759	45.658	13.726	41.251

注：同列数据栏中，经 Turkey 检验差异不显著的平均值之间用相同的字母表示（$P>0.05$）。

(Jiménez 和 Speisky, 2000; Jiménez 等, 2002)。而且有研究报道, GSH 的半胱氨基酸部分的巯基对 Cu、Hg、镉、铅具有较高的亲和力, 形成具有较高稳定能力的硫醇盐复合物, 便于机体排出体外 (Wang 和 Ballatori, 1998)。本实验研究发现, 添加 LA 的实验组的肝胰脏 GSH 和 MT 的含量均显著高于对照组, 推测 GSH 和 MT 在降低组织含量中发挥重要作用。

第四节 饲料中的硫辛酸对铜胁迫下皱纹盘鲍肝胰脏抗氧化指标的影响

(一) 对抗氧化指标的影响

如表 4-4 所示, 各 LA 处理组中皱纹盘鲍肝胰脏 SOD、CAT、Se-GPx、GST 的活力和 GSH 的含量存在显著性差异。700mg/kg LA 实验组的 SOD 活力显著高于对照组, 2 100mg/kg LA 实验组 SOD 高于对照组, 但与对照组无显著性差异。700、2 100mg/kg LA 实验组的 CAT、Se-GPx 和 GST 活力显著高于对照组。700mg/kg LA 实验组的 GSH 含量显著高于对照组。2 100mg/kg LA 实验组的 GSH 含量高于对照组, 但与对照组无显著性差异。各 LA 添加组中皱纹盘鲍肝胰脏的 TrxR、TrxP 的活性及 Trx 的含量无显著性差异。

研究表明, Cu 离子的氧化胁迫可以导致活性氧 (ROS) 的形成, 从而降低机体的抗氧化能力, 导致机体的氧化应激 (Company 等, 2004; Tamás 等, 2009)。LA 的抗氧化性被广泛认可, 其抗氧化性的研究主要表现在四个方面: 清除生物体内过量的活性氧, 保护机体免受氧化损伤; 可以再生内源性抗氧化剂; 修复氧化损伤。除此之外, 它还具有螯合金属的能力 (Gerreke 等, 1997; Lodge 等, 1998)。在本实验中发现, 在 Cu 离子的胁迫下, 饲料中添加 LA 显著提高皱纹盘鲍肝胰脏抗氧化酶 SOD、CAT、Se-GPx、GST 及 GSH 的含量, 表明 LA 可以提高 Cu 胁迫下皱纹盘鲍的抗氧化水平。在正常生理状况下, 机体的营养状态与自由基之间一直保持着动态的平衡, 但是如果该平衡受到破坏, 就会导致向

第四章 饲料硫辛酸解除铜对皱纹盘鲍毒性作用的研究

表 4-4 饲料中的硫辛酸（LA）对 Cu 胁迫下的皱纹盘鲍肝胰脏抗氧化相关指标的影响（平均值±标准误，$n=3$）

饲料中 LA 含量 (mg/kg)	SOD	CAT	Se-GPx	GST	GSH	Trx	TrxR	TrxP
0	7.34±0.08^b	4.40±0.12^b	12.15±0.16^c	2.21±0.04^c	4.86±0.38^b	296.20±1.12	2.50±0.02	4.62±0.13
700	11.81±0.88^a	6.69±0.59^a	20.38±0.44^a	2.84±0.03^a	6.50±0.27^a	294.68±0.47	2.57±0.06	4.44±0.26
2100	9.89±0.58^{ab}	7.09±0.47^a	14.35±0.70^b	2.59±0.03^b	5.67±0.10^{ab}	294.83±1.32	2.52±0.06	4.56±0.14
One-way ANOVA								
P 值	0.006	0.010	0.000	0.000	0.015	0.000	0.001	0.788
F 值	13.473	10.707	76.425	24.344	9.103	22.419	14.822	0.248

注：SOD，超氧化物歧化酶（U，每毫克蛋白中）；CAT，过氧化氢酶（U，每毫克蛋白中）；GPx，谷胱甘肽过氧化物酶（U，每毫克蛋白中）；GST，谷胱甘肽硫转移酶（U，每毫克蛋白中）；GSH，还原性谷胱甘肽（mg，每毫克蛋白中）；Trx，硫氧还蛋白（ng/mL）；TrxR，硫氧还蛋白还原酶（mU/L）；TrxP，硫氧还蛋白过氧化物酶（U，每毫克蛋白中）。同列数据栏中，经 Turkey 检验差异不显著的平均值之间用相同的字母表示（$P>0.05$）。

自由基的一方产生倾斜，从而造成机体细胞的氧化损伤。机体中的抗氧化酶系统在维持氧自由基平衡方面起着非常重要的作用（Palacea 等，1998），SOD、CAT、GPx、GST 和 GSH 是贝类内抗氧化酶的重要组分。SOD 是金属酶，含 Cu、Zn 或 Mn 原子，SOD 可以催化超氧阴离子产生 H_2O_2 和水，抑制了超氧阴离子向·OH 的转化，是抵抗氧自由基的第一个酶（Kappus，1985；Ruas 等，2008）。CAT 是过氧化物酶的标志性酶，可以催化 H_2O_2 转化成无害的水和氧气。本实验研究还发现，谷胱甘肽抗氧化系统在 LA 抗 Cu 的过程中发挥了重要作用。GSH 是体内衡量抗氧化能力的重要指标，在哺乳动物的研究中证明，外源 LA 可以再生机体 GSH（Busse 等，1992；Eason 等，2002）。Se-GPx、GST 都是以 GSH 为底物，可以清除机体的有机过氧化物和过氧化氢，一般情况下 Se-GPx 的作用比 GST 大。但也有研究表明，当 Se-GPx 活性下降时，GST 可以起补偿作用（Bell 等，1986）。根据刘学忠等（2004）等研究表明，LA 可以穿过血脑屏障，提高组织中 GPx 活性。GPx 的重要功能就是能够清除体内各种有机过氧化物和 H_2O_2，保护细胞免受自由基的毒害。小鼠中的研究也表明，外源补充 LA 可以明显减少组织脂质过氧化水平，升高心肌和肝脏 GSH 含量和 GPx、GST、GR 活性，提高机体抗氧化能力（李奕，2006）。

（二）对氧化指标的影响

如表 4-5 所示，700、2 100mg/kg LA 实验组的肝胰脏 MDA 的含量显著低于对照组。2 100mg/kg LA 实验组的蛋白羰基的含量显著低于对照组和 700mg/kg 组，且 700mg/kg LA 实验组蛋白羰基含量和对照组无显著性差异。2 100mg/kg LA 实验组 DNA 断裂程度显著低于对照组（$P<0.05$），700mg/kg 实验组 DNA 断裂程度与对照组无显著性差异。

Cu 可以通过 Fenton 氧化还原循环反应产生 ROS（Halliwell 和 Gutteridge，1984），并且可以参与脂质过氧化的激活和连锁反应（Viarengo 等，1990）。ROS 导致很多非酶蛋白的改变，其中蛋白羰基化被广泛用作氧化应激的标志性指标（Shacter 等，1994；Je 等，

表 4-5 饲料中的硒（Se）对 Cu 胁迫下皱纹盘鲍肝胰脏氧化指标的影响（平均值±标准误，$n=3$）

饲料中硒含量 (mg/kg)	MDA (nmol，每毫克蛋白中)	蛋白质羰基化 (nmo，每毫克蛋白中)	DNA 断裂程度（F 值）
对照	28.37 ± 0.22^a	7.99 ± 0.39^a	0.12 ± 0.02^b
700	25.33 ± 0.07^b	7.87 ± 0.63^a	0.25 ± 0.06^{ab}
2 100	25.86 ± 0.51^b	5.04 ± 0.77^b	0.37 ± 0.05^a
One-way ANOVA			
P 值	0.001	0.025	0.023
F 值	25.103	7.31	7.591

注：同列数据栏中，经 Turkey 检验差异不显著的平均值之间用相同的字母表示（$P>0.05$）。

2008)。Cu 的胁迫也可以导致 DNA 的损伤（Gabbianelli 等，2003）。本实验研究发现，700、2 100mg/kg 的 LA 实验组降低了 Cu 胁迫下皱纹盘鲍肝胰脏脂质过氧化水平及蛋白羰基化的水平，降低了 DNA 的断裂程度。饲料中添加硫辛酸有效降低了 Cu 胁迫下的氧化应激。原因首先依赖于 LA 较强的抗氧化能力，其次 LA（Devasagayam 等，1993；Seott 等，1994）和它的代谢产物 DHLA（Bonomi 等，1989；Müller 等，1990）能够螯合铁、铜、锰、铬、锌等过渡金属，降低·OH 的产生，阻断脂质过氧化，防止重金属中毒。

（三）对 MT 含量、MT mRNA 和 MTF-1 mRNA 表达的影响

如图 4-1 所示，700、2 100mg/kg 实验组的 MT 含量显著高于对照组，且 700mg/kg LA 实验组的 MT 含量显著高于 2 100mg/kg。700mg/kg 实验组的 MT mRNA 表达显著高于对照组，2 100 mg/kg LA 实验组的 MT mRNA 的表达高于对照组，但与对照组无显著性差异。700、2 100mg/kg LA 实验组的 MTF-1 mRNA 的表达均显著高于对照组。

抗氧化剂对皱纹盘鲍重金属的解毒作用研究

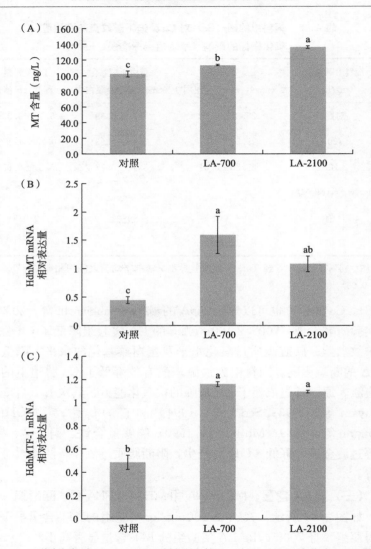

图 4-1　不同硫辛酸（LA）处理对皱纹盘鲍 Cu 胁迫下的肝胰腺中 MT 含量、HdhMT 和 HdhMTF-1 基因表达的影响（平均值±标准误，$n=3$）

（A）不同浓度下的 LA 对 Cu 胁迫下皱纹盘鲍肝胰脏 MT 含量的影响；（B）不同浓度下的 LA 对 Cu 胁迫下皱纹盘鲍肝胰脏 MT mRNA 表达的影响；（C）不同浓度下的 LA 对 Cu 胁迫下皱纹盘鲍肝胰脏 MTF-1 mRNA 表达的影响。

MT在抗重金属的应激中发挥重要的作用。首先，它可以通过巯基和自由基进行结合，降低重金属过程中产生的过量的ROS。其次，MT可与多种重金属结合，减轻重金属对生物体的氧化损伤。另外，MT中的巯基基团可以通过氢供体使受损的DNA得以修复。本实验研究发现，饲料添加700、2 100mg/kg LA可以升高皱纹盘鲍肝胰脏MT的含量及MT mRNA的表达。另外，本实验还发现MT的调控因子MTF-1的mRNA的水平随着MT mRNA的升高而显著升高，可以推测金属硫蛋白的表达受MTF-1的调控，说明MT的这一调控路径在LA抗Cu的氧化胁迫及降低重金属的含量中发挥重要的作用，但是具体的机制还需要进一步的研究。

小结：

（1）饲料中添加700、2 100mg/kg的硫辛酸可以显著降低皱纹盘鲍肝胰脏、血清、鳃、外套膜和肌肉组织中的Cu含量，并可以降低皱纹盘鲍肝胰脏中脂质过氧化的水平和蛋白羰基化水平及DNA的损伤，说明饲料中添加硫辛酸可以减轻Cu对皱纹盘鲍导致的氧化损伤。

（2）饲料中添加700、2 100mg/kg的硫辛酸显著升高了皱纹盘鲍肝胰脏抗氧化酶SOD、CAT、Se-GPx、GST的活性及GSH的含量，说明硫辛酸在一定程度上可以提高Cu胁迫下的皱纹盘鲍肝胰脏的抗氧化水平。

（3）饲料中添加硫辛酸显著升高了Cu胁迫下的皱纹盘鲍肝胰脏金属硫蛋白的含量及表达，而且显著升高了MTF-1 mRNA的表达。这表明在硫辛酸缓解Cu毒性的过程中，MT发挥了重要作用，具体机制还需要继续研究。

参 考 文 献

李奕, 2006. α硫辛酸对耐力训练小鼠力竭运动后谷胱甘肽抗氧化系统的影响 [D]. 上海：华东师范大学.

刘学忠, 崔旭, 卞建春, 等, 2004. 硫辛酸在大鼠全脑缺血再灌注损伤中的神

经保护作用 [J]. 中国兽医学报, 24 (4): 388-390.

田芳, 仲伟鉴, 应贤平, 等, 2007. α-硫辛酸对 H_2O_2 诱导的细胞活性氧水平及 DNA 氧化损伤的影响 [J]. 环境与职业医学, 24 (2): 180-189.

ARIVAZHAGAN P, RAMANATHAN K, PANNEERSELVAM C, 2001. Effect of DL-lipoic acid on mitochondrial enzymes in aged rats [J]. Chem. Biol. Interact, 138 (2): 189-198.

BELL JG, ADRON JW, COWEY CB, 1986. Effect of selenium deficiency on hydroperoxide-stimulated release of glutathione from isolated perfused liver of rainbow trout (*Salmo gairdneri*) [J]. Br. J. Nutr, 56: 421-428.

BONOMI F, CERIOLI A, PAGANI S, 1989. Molecular aspects of the removal of the removal of ferritin-bound iron by DL-dihydrolipoate [J]. Biochim. Biophys. Acta, 994 (2): 180-186.

BUSSE E, ZIMMER G, SCHOPOHL B, 1992. Influence of α-lipoic acid on intracellular glutathione in vitro and in vivo [J]. Arzneimttel-Forschung, 42 (6): 829-831.

CAO Z, TSANG M, ZHAO H, et al, 2003. Induction of endogenous antioxidants and Phase 2 enzymes by a-lipoic acid in rat cardiac H9C2 cells: protection against oxidative injure [J]. Biochem. Biophys. Res. Commun, 310: 979-985.

CAREY A, WILLIAMS, RHONDA M, et al, 2002. Lipoic acid as an antioxidant in mature thoroughbred geldings: A Preliminary Study [J]. J. Nutr, 2: 1628S-1631S.

COMPANY R, SERAFIM A, BEBIANNO MJ, et al, 2004. Effect of cadmium, copper and mercury on antioxidant enzyme activities and lipid peroxidation in the gills of the hydrothermal vent mussel *Bathymodiolus azoricus* [J]. Mar. Environ. Res, 58: 377-381.

DEVASAGAYAM TPA, SUBRAMANIAN M, PRADHAN DS, et al, 1993. Prevention of singlet oxygen-induced DNA damage by lipoate [J]. Chem. Biol. Interact, 86 (1): 79-92.

EASON RC, ARCHER HE, AKHTAR S, et al, 2002. Lipoic acid increases glucose uptake by skeletal muscles of obese-diabetic ob/ob mice [J]. Diabetes Obes. Metab, 4 (1): 29-35.

GABBIANELLI R, LUPIDI G, VILLARINI M, et al, 2003. DNA damage induced by copper on erythrocytes of gilthead sea bream*Sparusaurata* and mollusk *Scapharcainaequivalvis* [J]. Arch. Environ. Contam. Toxicol, 45: 350-356.

GERREKE PH, BIEWENGA, GUIDO R, 1997. The pharmacology of the antioxidant lipoic acid [J]. Gen. Pharmac, 29 (3): 315-331.

HALLIWELL B, GUTTERIDGE MC, 1984. Oxygen toxicity, oxygen radicals, transition metals and disease [J]. Biochem. J, 219: 1-14.

HARRIS ED, 1991. Copper transport: an overview [J]. Exp. Biol. Med, 196 (2): 130-140.

JE JH, LEE TH, KIM DH, et al, 2008. Mitochondrial ATP synthase is a target for TNBS-induced protein carbonylation in XS-106 dendritic cells [J]. Proteomics, 8: 2384-2393.

JIMÉNEZ I, ARACENA P, LETELIER ME, 2002. Chronic exposure of HepG2 cells to excess copper results in depletion of glutathione and induction of metallothionein [J]. Toxicology in vitro, 16 (2): 167-175.

JIMÉNEZ I, SPEISKY H, 2000. Effects of copper ions on the free radicalscavenging properties of reduced glutathione: implications of a complex formation [J]. Trace Elem. Biol. Med, 14: 161-167.

KAPPUS H, 1985. Lipid peroxidation: mechanisms, analysis, enzymology and biological relevance [M]. //SIESH. Oxidative Stress. London: Academic Press.

KRAMER K, HOPPE P P, PACKER L, 2001. Nutraceuticals in health and disease prevention [M]. New York: Marcel Dekker In C.

KOWLURU RA, ODENBACH S, BASAK S, 2005. Long-term administration of lipoic acid inhibits retinopathy in diabetic rats via regulating mitochondrial superoxide dismutase [J]. Invest Ophthalmol. Vis. Sci, 46: 396-422.

LEONARD P, RYBAK, KAZIM HUSAIN, 1999. Dose dependent protection by lipoic acid against cisplatin-induced ototoxicity in rats: antioxidant defense system [J]. Toxicol. SCI, 47 (3): 195-202.

LODGE JK, TRABER MG, PACKER L, 1998. Thiol chelation of Cu^{2+} by dihydrolipoic acid prevents human low density lipoprotein peroxidation [J]. Free Rad. Med, 25 (3): 287-297.

MAI K, ZHANG W, TAN B, et al, 2003. Effects of dietary zinc on the shell biomineralization in abalone *Haliotis discus hannai* Ino [J]. J. Exp. Mar. Biol. Ecol, 283: 51-62.

MONSERRAT JM, LIMA JV, FERREIRA JLR, et al, 2008. Modulation of antioxidant and detoxification responses mediated by lipoic acid in the fish *Corydoras paleatus* (Callychthyidae) [J]. Comp. Biochem. Physiol, Part C, 148: 287-292.

NAVARI IZZO F, QUARTACCI MF, SGHERRIC C. LIPOIC, 2002. Acid: a unique antioxidant in the detoxification of activated oxygen species. Plant Physiol. Biochem, 40: 463-470.

PACKER L, KRAEMER K, RIMBACH G, 2001. Molecular aspects of lipoic acid in the prevention of diabetes complications [J]. Nutrition, 17 (10): 888-895.

PACKER L, WITT EH, TRITSCHLER HJ, 1995. Alpha-lipoic acid as a biological antioxidant [J]. Free Radic. Biol. Med, 19 (2): 227-250.

PALACEA VP, BROWN SB, BARON CI, et al, 1998. An evaluation of the relationships among oxidative stress, antioxidant vitam ins and early mortality syndrome (EMS) of lake trout (*Salvelinus namaycush*) from Lake Ontario [J]. Aquat. Toxicol, 43: 259-268.

QUINN JF, BUSSIERE JR, HANLLNOND RS, et al, 2007. Chronic dietary α-lipoic acid reduces deficits in hippocampal memory of aged Tg 2576 mice [J]. Neurobiol. Aging, 28: 213-225.

REED, LESTER J, 2001. A trail of research from lipoic acid to alpha-keto acid dehydrogenase complexes [J]. J. Biol. Chem, 276 (42): 38329-38336.

RICHARDS M, RECENT, 1989. Recent developments in trace element metabolism and function: role of metallthionein in copper and zinc metabolism. J. Nutr, 119: 1062-1070.

RUAS CBG, CARVALHO CS, DE Araujo HSS, 2008. Oxidative stress biomarkers of exposure in the blood of cichlid species from a metal-contaminated river [J]. Ecotoxicol. Environ. Saf, 71: 86-93.

SHACTER E, WILLIAMS JA, LIM M, et al, 1994. Differential susceptibility of plasma proteins to oxidative modification: examination by western blot immunoassay [J]. Free Radical Biol. Med, 17: 429-437.

TAMÁS L, VALENTOVICOVÁ K, HALUSKOVÁ L, et al, 2009. Effect of cadmium on the distribution of hydroxyl radical, superoxide and hydrogen peroxide in barley root tip [J]. Protoplasma, 236: 67-72.

THORNALLEY PJ, VASAK M, 1985. Possible role for metallothionein in protection against radiation-induced oxidative stress. Kinetics and mechanism of its reaction with superoxide and hydroxyl radicals [J]. Biochim. Biophys. Acta, 827: 36-44.

VIARENGO A, CANESI L, PERTICA M, 1990. Heavy metal effect on lipid peroxidation in the tissues of *Mytilus galloprovincialis* L [J]. Comp. Biochem. Physiol (C), 97: 37-42.

WANG HW, XU HM, XIAO GH, 2010. Effects of selenium on the antioxidant enzymes response of Neocaridinaheteropodaexposed to ambient nitrite. Bull [J]. Environ. Contam. Toxicol, 84: 112-117.

WANG W, BALLATORI N, 1998. Endogenous glutathione conjugates: occurrence and biological functions. Pharmacol Rev 50: 335-356 [J]. Pharmacol. Rev, 50 (3): 335-356.

ZHANG L, XING GQ, BARKER JL, 2001. α-lipoic acid protects rat cortical neurons against cell death indueed by amyloid and hydrogen peroxide through the Akt signaling pathway [J]. Neurosci. Lett, 312 (3): 125-128.

ZHANG W, MAI K, XU W, et al, 2007. Interaction between vitamins A and D on growth and metabolic responses of abalone *Haliotis discus hannai*, Ino [J]. J. Shellfish Res, 26: 51-58.

第五章
水中镉对皱纹盘鲍抗氧化反应、脂质过氧化和体内金属沉积的影响

第一节 引 言

在世界范围内，重金属镉（Cd）是出现在许多沿海地区的最常见的污染物。由于其毒性、来源广泛、生物积累性和难降解性等特点，重金属 Cd 已构成相当大的环境问题。在中国，随着工业化的快速发展，化石燃料燃烧、垃圾焚烧、工业废物的排放等造成了普遍的 Cd 污染，这些污染严重威胁着水生生态系统。Cd 具有很强的致癌性。我国的锦州湾、渤海湾、胶州湾的水中 Cd 含量超过了国家《渔业水质标准》（GB 11607—1989）（Xu 等，2000；Xu 等，2005；Chen 等，2004）。Cd 的半衰期很长，大量的 Cd 离子进入水体中可以引起水生生物的中毒，对其器官、系统等产生毒性影响，造成水生生物的畸变甚至死亡（郭永灿等，1989）。

大量的实验研究证明，Cd 对生物体产生毒性效应的主要原因是造成机体的氧化损伤（刘晓玲等，2003）。Cd 虽然不像 Cu 那样可以直接诱发活性氧，但可以作用于机体的抗氧化酶系统，导致酶的表达与活力发生变化。体内的抗氧化酶系统主要包括超氧化物歧化酶（SOD）、过氧化氢酶（CAT）、谷胱甘肽过氧化物酶（GPx）、谷胱甘肽硫转移酶（GST）和抗氧化小分子 GSH 等。这些酶类和抗氧化的小分子可以积极有效地清除体内的活性氧，从而避免活性氧对机体

造成氧化损伤。活性氧可以对生物体的多种大分子造成损害,可以使脂类中的不饱和脂肪酸氧化,从而产生脂质过氧化物,脂质过氧化物进一步分解成丙二醛,它们甚至使蛋白质氧化或 DNA 发生断裂等(刘瑞明等,1990)。

海水的贝类因为蓄积这些污染物的能力,所以很容易受到环境污染物的影响,从而加剧了 ROS 的产生(Company 等,2004;Company 等,2006)。这些生物标志物的活性变化可以间接反映环境氧化应激的存在,因此常作为环境污染胁迫的敏感指标(Lemairon 等,1994)。随着水体污染的加剧,研究活性标志物的变化作为检测环境胁迫的标志具有非常重要的意义(Powera 和 Chapman,1992;Chapman,1995)。皱纹盘鲍是海洋大型原始腹足类,是中国北方海水养殖贝类的主要品种之一。本书重点研究了不同梯度的镉对皱纹盘鲍毒性作用,以抗氧化酶、抗氧化小分子(如 GSH),以及肝脏脂质过氧化等生物指标为依据,为利用皱纹盘鲍进行生物监测、防治水体重金属污染提供科学依据。这样既可以加强鲍的养殖管理,又可以提高水产动物的食物安全,为找到合适的解毒途径提供条件。

第二节 摄食生长实验设计

(一)实验动物与药品

皱纹盘鲍幼鲍和新鲜的海带均购买于青岛鳌山卫育苗场,所购买的鱼苗为同一批鲍苗。在中国海洋大学水产馆养殖系统中暂养 2 周。暂养期间,每天 18:00 饱食投喂海带一次,次日 8:00 清底换水,养殖过程中密切观察采食及健康状况。

$CdCl_2 \cdot 2.5H_2O$ (AR)购自广州化学试剂厂,实验前,先用蒸馏水配制成 Cd 浓度为 1 000mg/L 的母液,再在 5 个 350L 实验用洁净桶中用天然海水稀释成实验所需要的各种浓度,然后用水泵抽到各个玻璃缸中,玻璃缸的体积为 100L。硝酸、硫酸、过氧化氢、盐酸均为优级纯(GR)。

（二）摄食实验设计

根据直线内插法得到的重金属 Cd 对皱纹盘鲍的半致死浓度为 2.848mg/L，且累计死亡率（y）和浓度（x）的相关直线方程为：$y=0.314x-0.395$（$R^2=0.909$）。根据 Cd 的半致死浓度和国家渔业水质标准（Cd 含量≤0.005mg/L），进而确定 Cd 的亚慢性毒性实验浓度。实验设 0（对照组）、0.025mg/L、0.05mg/L、0.25mg/L、0.5mg/L 共 5 个处理组，每个处理组均设 3 个重复。海水中 Cd 的实测浓度为：4.5μg/L、（0.027±0.00）mg/L、（0.050±0.00）mg/L、（0.27±0.01）mg/L、（0.53±0.02）mg/L。挑选规格一致的健康个体随机分成 15 组，每缸 60 只鲍。在整个实验过程中，保持水中溶氧量在 6mg/L 以上，pH 为 7.4～7.9，水温控制在 18～22℃，盐度为 22～28，光暗比为 12h：12h。实验时间为 28d，每天早晚换水两次，更换总量为各组溶液的 50%。每天 18：00 投喂新鲜海带一次，第二天 8：00 清除粪便和残饵，如有个体死亡，及时捞出并且计数。实验海水的水样每 2d 取一次以便以后实验分析。实验检测海带中 Cd 的含量为 4.45mg/kg。

第三节 镉浓度对皱纹盘鲍死亡率和抗氧化反应的影响

（一）对死亡率的影响

在各 Cd 离子浓度下，各实验组和对照组的皱纹盘鲍的存活率均为 100%。

（二）Cd 浓度对抗氧化酶和抗氧化小分子谷胱甘肽的影响

在 Cd 离子的胁迫下，皱纹盘鲍肝胰脏 SOD、CAT、GPx、GST 的活性及 GSH 的含量的变化见表 5-1。

总体来说，在各 Cd 浓度下，随着时间延长，肝胰脏 SOD 的活力基本上呈现先显著性升高后降低到对照水平又升高的过程。在实验的第 6 天，各浓度肝胰脏 SOD 的活力均显著高于对照组。在实验的第 10 天和第 15 天，各实验组 SOD 的活力和对照组无显著性

第五章 水中镉对皱纹盘鲍抗氧化反应、脂质过氧化和体内金属沉积的影响

表 5-1 水体中 Cd 对皱纹盘鲍肝胰脏抗氧化相关指标的影响（平均值±标准误，$n=3$）

指标	Cd浓度	第0天	第1天	第3天	第6天	第10天	第15天	第21天	第28天
SOD	对照	29.66±0.55	30.12±1.26^{AB}	28.74±0.39^{B}	28.53±0.44^{C}	27.82±1.15	29.42±0.29	27.38±1.36^{B}	29.69±0.82^{B}
	0.025mg/L	29.66±0.55^{ab}	27.65±1.34^{ABab}	33.74±2.30^{ABab}	35.59±1.12^{Aa}	28.36±3.68^{ab}	30.55±5.10^{ab}	22.11±1.31^{Cb}	29.88±1.37^{Aab}
	0.05mg/L	29.66±0.55^{cd}	32.21±1.16^{Abcd}	38.88±1.45^{Aa}	31.84±0.27^{Bcd}	28.61±0.84^{d}	33.53±1.17^{bc}	27.36±1.08^{Bd}	36.93±0.97^{Aa}
	0.25mg/L	29.66±0.55^{cd}	25.52±1.46^{Bd}	31.70±1.73^{Bc}	33.86±0.05^{ABbc}	25.89±0.39^{d}	33.14±0.53^{abc}	36.44±0.72^{ab}	37.38±1.27^{Aa}
	0.5mg/L	29.66±0.55^{bc}	26.01±1.65^{ABd}	31.03±0.06^{Bbc}	32.61±0.36^{Bb}	29.07±0.10^{cd}	32.45±0.32^{b}	36.92±0.02^{Aa}	36.18±1.60^{Aa}
CAT	对照	10.78±0.40	10.81±0.66	10.66±1.18	11.95±0.13^{D}	10.47±0.87^{B}	9.52±0.25^{AB}	9.86±0.30^{D}	11.33±0.82^{D}
	0.025mg/L	10.78±0.40^{c}	11.24±0.56^{c}	11.22±1.52^{c}	21.79±0.76^{Cb}	12.23±1.33^{Bc}	8.90±1.41^{ABc}	30.42±0.54^{Aa}	26.74±1.41^{Cab}
	0.05mg/L	10.78±0.40^{d}	9.67±0.74^{d}	9.34±0.18^{d}	31.67±0.69^{Ad}	21.82±1.58^{Ad}	12.33±1.52^{Ad}	33.27±0.80^{Ab}	43.09±0.87^{Aa}
	0.25mg/L	10.78±0.40^{d}	8.58±1.11^{d}	11.30±0.32^{d}	23.24±1.15^{BCc}	20.29±0.95^{Abc}	9.63±1.25^{ABd}	17.91±1.15^{Cc}	31.33±0.42^{Ba}
	0.5mg/L	10.78±0.40^{c}	9.27±0.56^{c}	10.56±1.52^{c}	26.34±0.76^{Bb}	22.86±1.33^{Ab}	6.80±1.41^{Bc}	26.01±0.54^{Bb}	35.15±1.63^{Bab}
GPx	对照	50.75±2.58	46.16±7.45	46.46±4.60	50.84±3.33^{A}	48.66±2.63^{A}	52.81±3.97^{A}	58.70±4.56^{A}	52.13±4.19^{AB}
	0.025mg/L	50.75±2.58^{bc}	47.57±0.18^{abc}	53.14±6.97^{ab}	59.11±2.07^{Aa}	30.06±0.73^{Bc}	42.35±2.67^{ABabc}	39.12±1.23^{Bbc}	46.80±5.87^{ABabc}
	0.05mg/L	50.75±2.58^{ab}	45.59±0.41^{ab}	48.23±4.98^{ab}	48.53±0.47^{ABab}	29.52±1.92^{Bb}	27.64±0.97^{Cc}	40.17±3.44^{Bbc}	56.96±5.31^{Aa}
	0.25mg/L	50.75±2.58^{ab}	42.18±2.10^{bc}	61.86±3.80^{a}	37.86±3.16^{Bc}	31.23±1.61^{Bc}	34.57±3.08^{BCc}	41.19±4.06^{ABbc}	33.26±0.17^{Cc}
	0.5mg/L	50.75±2.58^{a}	39.48±1.68^{abc}	49.60±4.30^{ab}	33.54±2.43^{Cc}	27.46±2.96^{Bc}	35.15±1.71^{BCbc}	35.72±3.37^{Bbc}	37.04±3.96^{BCbc}
GST	对照	48.38±0.55	57.01±3.37	56.13±1.18^{C}	55.19±0.34^{AB}	54.56±1.52	54.22±4.82	51.98±4.10	53.13±1.76^{B}
	0.025mg/L	48.38±0.55^{b}	55.20±1.99^{ab}	38.52±1.52^{BCab}	48.73±1.99^{Bab}	57.87±1.84^{ab}	56.24±2.37^{ab}	51.02±1.67^{ab}	60.11±3.20^{ABa}

(续)

指标	Cd浓度	第0天	第1天	第3天	第6天	第10天	第15天	第21天	第28天
GST	0.05mg/L	48.38±0.55^{d}	57.18±4.07^{bc}	64.74±3.05^{Bb}	41.30±0.42^{cd}	49.85±2.47^{cd}	52.23±1.12^{bcd}	48.79±2.46^{cd}	70.17±3.97^{Aa}
	0.25mg/L	48.38±0.55^{de}	54.29±2.71^{de}	78.09±3.01^{Aa}	41.25±3.46^{Bd}	53.22±3.71^{cd}	57.34±1.09^{bc}	58.01±1.34^{bc}	69.06±4.56^{Aab}
	0.5mg/L	48.38±0.55^{b}	49.65±3.18^{b}	61.46±3.41^{BCab}	50.12±0.77^{Ab}	55.77±4.74^{ab}	51.49±5.27^{bc}	54.52±5.24^{ab}	69.98±3.17^{Aa}
GSH	对照	4.47±0.04	4.07±0.20	4.19±0.27^{C}	4.23±0.08	3.92±0.19^{A}	3.50±0.45^{B}	3.90±0.32^{B}	4.37±0.32^{C}
	0.025mg/L	4.47±0.04^{ab}	3.81±0.12^{bc}	4.73±0.19^{BCa}	4.66±0.12^{a}	3.07±0.09^{Bcd}	2.77±0.32^{Bd}	2.46±0.15^{Cd}	2.70±0.13^{Cd}
	0.05mg/L	4.47±0.04^{b}	3.75±0.27^{bc}	5.72±0.06^{Aa}	4.49±0.19^{bc}	2.81±0.16^{Bd}	2.99±0.13^{Bd}	2.61±0.19^{Cd}	2.79±0.13^{Cd}
	0.25mg/L	4.47±0.04^{bc}	3.30±0.21^{c}	5.28±0.11^{ABb}	4.34±0.10^{bc}	3.88±0.24^{Abc}	4.38±0.28^{ABbc}	4.41±0.40^{Ab}	7.13±0.59^{Ba}
	0.5mg/L	4.47±0.04^{cde}	3.31±0.16^{bc}	5.14±0.23^{ABe}	4.69±0.15^{cd}	3.97±0.10^{Ade}	5.60±0.51^{Ac}	8.78±0.14^{Ab}	10.63±0.65^{Aa}
T-AOC	对照	2.07±0.08	2.32±0.09	2.39±0.12^{C}	2.43±0.19^{B}	2.29±0.15	2.55±0.03^{BC}	2.30±0.18^{C}	2.73±0.17^{C}
	0.025mg/L	2.07±0.08^{c}	2.48±0.03^{abc}	3.20±0.15^{Bab}	3.51±0.25^{Ba}	2.37±0.19^{bc}	2.10±0.17^{c}	1.72±0.24^{Cc}	2.64±0.44^{Cabc}
	0.05mg/L	2.07±0.08^{c}	2.71±0.23^{bc}	3.86±0.14^{Aa}	3.11±0.07^{ABab}	2.54±0.23^{bc}	2.75±0.03^{Bbc}	2.28±0.34^{BCbc}	3.78±0.06^{Bb}
	0.25mg/L	2.07±0.08^{c}	2.33±0.19^{bc}	3.27±0.14^{Bb}	3.15±0.25^{ABb}	2.70±0.05^{bc}	3.28±0.33^{ABb}	3.32±0.28^{Bb}	4.66±0.16^{Ba}
	0.5mg/L	2.07±0.08^{e}	2.67±0.15^{cd}	3.36±0.06^{ABbc}	3.18±0.22^{ABcd}	2.96±0.10^{cd}	4.03±0.12^{Ab}	5.53±0.22^{Aa}	6.12±0.02^{Aa}

注：CAT，过氧化氢酶（U，每毫克蛋白中）；SOD，超氧化物歧化酶（U，每毫克蛋白中）；GPx，谷胱甘肽过氧化物酶（U，每毫克蛋白中）；GST，谷胱甘肽硫转移酶（U，每毫克蛋白中）；GSH，还原性谷胱甘肽（mg，每克蛋白中）；T-AOC，总抗氧化力（mmol/mg）。同行数据中，经Tukey检验差异不显著的平均值之间用相同的小写字母表示（$P>0.05$）；同列数据中，经Tukey检验差异不显著的平均值之间用相同的大写字母表示（$P>0.05$）。

差异。在实验的第 21、28 天，在 0.25、0.5mg/L Cd 处理组 SOD 活力显著高于对照组。

在各浓度 Cd 下，皱纹盘鲍肝胰脏 CAT 的活力呈现先升高后降低到对照水平然后又升高的过程。在实验的第 6、21 和 28 天，各浓度处理下 CAT 的活力均显著高于对照组。

在 0.25、0.5mg/L Cd 浓度下，肝胰脏 GPx 的活性随着时间的延长大致呈下降的趋势，且从实验的第 6 天开始均显著低于对照组。在 0.025、0.5mg/L Cd 浓度下，GPx 的活性随着实验的延长先降低后升高，且在实验的第 10 天显著低于对照组。

在各个 Cd 浓度下，肝胰脏 GST 的活力随着时间的延长基本上呈现先升高后降低到对照水平后又升高的过程。在实验的第 3 天和第 28 天，各浓度下 GST 的活力均高于对照组。在实验的第 3 天，在 0.05、0.25mg/L Cd 浓度下，GST 的活力显著高于对照组。在实验的第 28 天，在 0.05、0.25、0.5mg/L Cd 浓度下，GST 的活力显著高于对照组。在实验的第 10~21 天，各浓度处理组的 GST 的活力和对照组无显著性差异。

总体来说，在 0.025、0.05mg/L Cd 浓度下，肝胰脏 GSH 的含量随着时间的延长呈先升高后降低到对照水平然后又降低的过程。在 0.25、0.5mg/L Cd 浓度下，GSH 的含量先升高后降低到对照水平然后又升高。在实验的第 3 天，0.05、0.25、0.5mg/L Cd 浓度下，肝胰脏 GSH 的含量显著高于对照组。

总体来说，在 0.025mg/L Cd 浓度下，T-AOC 的活力随着时间增加先升高后降低到对照水平，在 0.05、0.25、0.5mg/L Cd 胁迫下，T-AOC 的活力随着时间的延长先升高后降低到对照水平然后又升高。在实验的第 3 天，各个浓度 Cd 处理的 T-AOC 均显著高于对照组。

重金属 Cd 被证明可以导致 ROS 的产生（Di Giulio 等，1995；Tamás 等，2009），从而使水产贝类产生氧化应激（Company 等，2004）。机体中的抗氧化酶系统在维持氧自由基平衡方面起着非常重要的作用，抗氧化系统主要包括一些抗氧化酶

(SOD、CAT、GPx、GST) 和抗氧化的小分子 GSH。这些抗氧化成分可以通过积极的变化来消除过量的 ROS，以达到保护水产动物免受氧化应激的伤害 (Thomas 和 Wofford, 1993; Reméo 等, 2000; Asagba 等, 2008; Cao 等, 2010)。从本实验可以看出，抗氧化酶和抗氧化的小分子积极对抗 Cd 导致的危害，在保护机体免受 Cd 的氧化应激中发挥着重要作用。本研究发现，在 Cd 离子的胁迫下，肝胰脏 SOD、CAT、GST 的活力具有相似的变化趋势，它们均随着时间的延长先升高后降低到对照水平然后又升高。GPx 的活力随着时间的延长呈下降的趋势。SOD 的活性在实验第 1 天就发生了显著性变化，GST、GSH 和 T-AOC 在实验的第 3 天发生显著性变化，CAT 和 GPx 在实验的第 6 天发生了显著性变化。以上说明 SOD、CAT、GST 和 GSH 在抵抗 Cd 的氧化应激中发挥更大的作用，且 SOD 在抵抗 Cd 离子中最敏感。

本实验研究表明，Cd 胁迫显著诱导了皱纹盘鲍肝胰脏 SOD 活性的升高。Cd 可以导致 SOD 活性升高的现象在鱼类 (Pandey 等, 2003; Asagba 等, 2008) 和软体动物 (Doyotte 等, 1997; Regoli 等, 1998; Company 等, 2004, 2010) 中均被发现。说明在重金属 Cd 的胁迫下，皱纹盘鲍肝胰脏中产生了大量的超氧阴离子，SOD 积极参与超氧阴离子的分解过程，使机体免受自由基的损害。

CAT 是重要的抗氧化酶，它和 GPx 均可以催化 H_2O_2 分解成水。Warner (1994) 研究表明，SOD 的显著变化一定伴随着 CAT 或 GPx 的变化。本实验研究发现，肝胰脏 CAT 和 SOD 的活性显著性被激活，但是 GPx 的活性却显著性被抑制。这表明相比较 GPx 而言，CAT 在皱纹盘鲍抵抗 Cd 产生的活性氧中发挥更重要的作用。Livingstone 等 (1992) 研究表明，CAT 在无脊椎动物中发挥比 GPx 更重要的作用。Yu (1994) 研究发现，在低浓度的 H_2O_2 下，GPx 发挥重要作用，但是在高浓度的 H_2O_2 下，CAT 则发挥更重要的作用。由此可以推测在 Cd 胁迫下，皱纹盘鲍肝胰

脏产生了大量的 H_2O_2，从而使 CAT 被显著性诱导。

GSH 被认为是细胞防御中最重要的抗氧化剂（Meister 和 Anderson，1983；Segner 和 Braunbeck，1998）。这个富含巯基的三肽物质通过与金属离子螯合或改变金属的吸收来达到缓解金属毒性的作用（Foulkes，1993；Ochi 等，1988；Burton 等，1995）。在 Cd 胁迫下 GSH 升高的现象在很多鱼中被发现，如虹鳟、罗非鱼、鲻（Thomas 等，1982；Tort 等，1996；Firat 等，2009）。在本实验研究中，在 Cd 胁迫下，肝胰脏 GSH 呈现先升高后降低然后又变化的趋势，说明 GSH 在皱纹盘鲍抗 Cd 的氧化应激中发挥重要作用。

一些领域的研究表明，GST 在贻贝应对有机污染物中发挥重要作用（Moreira 和 Guilhermino，2005；Rocher 等，2006），且被认为是贻贝有机污染的标志性物质。在 28d 的 Cd 胁迫实验中，GST 在实验的第 3、28 天被显著性诱导。Fernández 等（2010）研究发现，在 Cd 污染严重的海域，贻贝具有较高的 GST 活性。以上研究表明，GST 在 Cd 的胁迫中起重要作用。

第四节　镉浓度对皱纹盘鲍脂质过氧化和体内金属沉积的影响

（一）对脂质过氧化的影响

如表 5-2 所示，在各个 Cd 浓度下，皱纹盘鲍肝胰脏脂质过氧化水平无显著性变化。

在本实验中，皱纹盘鲍肝胰脏和肌肉 Cd 的含量无论与时间还是浓度都具有良好的剂量关系。类似的 Cd 的积累模式在其他水产动物中也被发现，如美国鳗（Gill 等，1992）、罗非鱼（Pelgrom 等，1995）、虹鳟（McGeer 等，2000）中。实验还发现，皱纹盘鲍肝胰脏比肌肉积累更高的 Cd。这与其他研究结果相类似（Usha Rani 和 Ramamurthi，1989；Asagba 和 Obi，2001；Jayakumar 和 Paul，2006；Samuel 等，2008）。很多研究表明肝脏是代谢、隔离和排泄外来物

表 5-2 水体中的 Cd 浓度对皱纹盘鲍肝胰脏 MDA 含量的影响（平均值±标准误，$n=3$）

Cd 浓度	第 0 天	第 1 天	第 3 天	第 6 天	第 10 天	第 15 天	第 21 天	第 28 天
对照	4.93±0.12	5.92±0.31	5.56±0.46	4.99±0.49	4.62±0.26	5.58±0.11	4.73±0.40	6.23±0.38
0.025mg/L	4.93±0.12	5.23±0.20	5.45±1.01	8.79±1.08	4.95±0.57	5.70±1.03	6.68±0.26	6.92±0.77
0.05mg/L	4.93±0.12	4.45±0.44	6.50±1.00	5.10±0.85	4.53±0.42	5.56±0.86	5.94±0.19	8.36±1.18
0.25mg/L	4.93±0.12	3.73±0.37	5.33±1.18	5.12±0.57	4.65±0.68	6.40±0.42	7.63±1.20	7.20±0.85
0.5mg/L	4.93±0.12	4.26±0.43	5.92±0.15	6.02±0.13	4.35±0.26	5.36±0.26	6.36±0.98	5.84±0.27

注：MDA，丙二醛（nmol，每毫克蛋白中）。

同行数据中，经 Tukey 检验差异不显著的平均值之间用相同的大写字母表示（$P>0.05$）；同列数据中，经 Tukey 检验差异不显著的平均值之间用相同的小写字母表示（$P>0.05$）。

质的主要器官。Sorensen（1991）认为，肌肉中低的重金属积累的原因是在肌肉中重金属和巯基化合物的结合率低。软体动物除扇贝和牡蛎外，供人食用的 Cd 含量不应超过 1.0mg/kg（以鲜重计）（NY 5073—2006）。Cd 的积累在各浓度处理 28d 下，均高于国家规定水平。

（二）对组织中金属含量的影响

如表 5-3 所示，总体上来说，肝胰脏和肌肉的 Cd 含量随着水中 Cd 离子浓度的升高和时间的延长呈上升的趋势。肝胰脏和肌肉中 Cd 离子含量最高值为 282.96、10.4μg/g，分别是对照组的 70、10 000 倍。

重金属会通过脂质过氧化对细胞造成伤害，被认为是重金属对细胞产生危害的第一步（Viarengo，1989）。MDA 被认为是细胞膜脂质过氧化反应的主要活性产物（Ohkawa 等，1979）。一些研究表明，Cd 的胁迫导致软体动物脂质过氧化水平的提高（Geret 等，2002；Company 等，2004；Chandrana 等，2005）。但是根据 Viarengo（1990）研究发现，Cd 的胁迫并没有引起贻贝鳃和消化腺 MDA 的升高。

在本实验中，在各浓度 Cd 胁迫下，并没有引起 MDA 的显著性升高，这可能是抗氧化系统发挥的作用，抗氧化酶和抗氧化的小分子积极应对 Cd 的氧化胁迫，从而避免了机体脂质过氧化的发生。

小结：

（1）在各浓度 Cd 胁迫下，抗氧化酶 SOD、CAT、GST 和抗氧化的小分子 GSH 在对抗 Cd 的氧化胁迫中发挥重要的作用，且 SOD 为 Cd 胁迫最敏感的指标，其次为 GSH、GST 和 T-AOC。

（2）即使在 Cd 最高浓度 0.5mg/L（国家渔业水质标准的 100 倍），在实验的 28d 内也未造成皱纹盘鲍的死亡，且未对肝胰脏脂质过氧化水平造成影响。这可能与抗氧化酶的积极参与有关。

（3）肝胰脏和肌肉的 Cd 含量随着时间和浓度的增加呈上升的趋势，且肝胰脏 Cd 积累量高于肌肉。

表 5-3 水体中的 Cd 对皱纹盘鲍肝胰脏和肌肉 Cd 含量（$\mu g/g$）的影响（平均值±标准误，$n=3$）

组织	Cd 浓度	第 0 天	第 1 天	第 3 天	第 6 天	第 10 天	第 15 天	第 21 天	第 28 天
肝胰脏	对照	2.76±0.19	2.83±0.48[D]	2.42±0.23[D]	2.75±0.13[C]	2.99±0.20[D]	3.65±0.59[D]	3.09±0.27[C]	4.07±0.43[E]
	0.025mg/L	2.76±0.19[g]	2.23±0.14[CDg]	4.05±0.25[CDf]	5.86±0.25[Ce]	7.37±0.12[DCd]	10.23±0.11[CDc]	14.14±0.03[Cb]	21.13±0.22[Da]
	0.05mg/L	2.76±0.19[f]	4.29±0.04[Cf]	6.42±0.67[Cef]	10.09±0.39[Cde]	14.17±1.10[Cd]	19.73±1.27[Cc]	30.91±2.29[Cb]	39.42±1.48[Ca]
	0.25mg/L	2.76±0.19[e]	8.89±0.53[Be]	17.85±0.89[Be]	42.24±2.72[Bd]	60.77±0.72[Be]	63.61±7.07[Bc]	136.16±2.40[Bb]	151.46±3.57[Ba]
	0.5mg/L	2.76±0.19[e]	15.51±0.21[Ae]	29.07±1.53[Ae]	90.34±3.35[Ad]	128.35±1.15[Ac]	196.65±11.31[Ab]	252.37±16.62[Aab]	282.96±4.95[Aa]
肌肉	对照	0.00±0.00[b]	0.00±0.00[Cb]	0.00±0.00[Cb]	0.00±0.00[Cb]	0.00±0.00[Db]	0.00±0.00[Cb]	0.00±0.00[Cb]	0.00±0.00[Ea]
	0.025mg/L	0.00±0.00[c]	0.06±0.00[c]	0.04±0.00[Cc]	0.09±0.01[Bc]	0.19±0.04[Dc]	0.48±0.02[Cc]	0.46±0.01[Cb]	0.71±0.10[Da]
	0.05mg/L	0.00±0.00[d]	0.11±0.01[Cd]	0.11±0.01[Cd]	0.33±0.06[Bcd]	0.85±0.00[Cbc]	0.75±0.07[Cb]	1.59±0.26[Ca]	1.59±0.13[Ca]
	0.25mg/L	0.00±0.00[d]	0.61±0.01[Bcd]	0.62±0.08[Bcd]	1.74±0.04[Ac]	3.16±0.17[Bc]	4.27±0.60[Bab]	4.89±0.36[Bb]	5.50±0.14[Ba]
	0.5mg/L	0.00±0.00[f]	0.81±0.05[Ae]	0.97±0.06[Ae]	2.29±0.26[Ad]	4.18±0.11[Ac]	8.06±0.06[Ac]	9.97±0.18[Ab]	10.40±0.17[Aa]

注：同行数据中，经 Tukey 检验差异不显著的平均值之间用相同的小写字母表示（$P>0.05$）；同列数据中，经 Tukey 检验差异不显著的平均值之间用相同的大写字母表示（$P>0.05$）。

第五章 水中镉对皱纹盘鲍抗氧化反应、脂质过氧化和体内金属沉积的影响

参 考 文 献

刘瑞明，刘毓谷，1990. 镉的肝细胞毒性与脂质过氧化关系的研究 [J]. 中国环境科学学报 (103)：187-191.

刘晓玲，周忠良，陈立侨，2003. 镉对中华绒螯蟹 (*Eriocheir sinensis*) 抗氧化酶活性的影响 [J]. 海洋科学，27 (8)：59-62.

郭永灿，周青山，姚正辉，1989. 镉对白鲢肾组织细胞的纤维结构及超微结构损伤的影响 [J]. 中国环境科学，9 (2)：117-122.

ASAGBA SO, ERIYAMREMU GE, IGBERAESE ME, et al, 2008. Bioaccumulation of cadmium and its biochemical effect on selected tissues of the catfish (*Clarias gariepinus*) [J]. Fish Physiol. Biochem, 34：61-69.

ASAGBA SO, OBI FO, 2001. Cadmium uptake: dose and time dependent tissue load and redistribution in the catfish, *Clarias anguilaris* (Line, 1758) [J]. Trop J. Environ. Sci. Health, 4 (1)：1-6.

BURTON CA, HATLELID K, DIVINE K, 1995. Glutathione effects on toxicity and uptake of mercuric chloride and sodium arsenite in rabbit renal cortical slices [J]. Environ. Health Perspect, 103 (1)：81-84.

CAO L, HUANG W, LIU J, et al, 2010. Accumulation and oxidative stress biomarkers in Japanese flounder larvae and juveniles under chronic cadmium exposure [J]. Comp. Biochem. Physiol. C, 151：386-392.

CHANDRANA R, SIVAKUMARA AA, MOHANDASSB S, et al, 2005. Effect of cadmium and zinc on antioxidant enzyme activity in the gastropod, *Achatina fulica* [J]. Comp. Biochem. Physiol. C, 140：422-426.

CHAPMAN PM, 1995. Sediment quality assessment: status and outlook [J]. J. Aquat. Ecosyst. Health, 4：183-194.

CHEN JL, LIU WX, LIU SZ, et al, 2004. An evaluation on heavy metal contamination in the surface sediments in Bohai SeaMar [J]. Sci, 28：16-21.

COMPANY R, FELI'CIA H, SERAFIMA A, et al, 2010. Metal concentrations and metallothionein-like protein levels in deep-sea fishes captured near hydrothermal vents in the Mid-Atlantic Ridge off Azores [J]. Deep-Sea Research I, 57：893-908.

COMPANY R, SERAFIM A, BEBIANNO MJ, et al, 2004. Effect of cadmium, copper and mercury on antioxidant enzyme activities and lipid peroxidation in the gills of the hydrothermal vent mussel *Bathymodiolus azoricus* [J]. Mar. Environ. Res, 58: 377-381.

COMPANY R, SERAFIM A, COSSON R, et al, 2006. The effect of cadmium on antioxidant responses and the susceptibility to oxidative stress in the hydrothermal vent mussel *Bathymodiolus azoricus* [J]. Mar. Biol, 148: 817-825.

DI GIULIO R, BENSON W, SANDERS B, et al, 1995. Biochemical mechanisms: metabolism, adaptation and toxicity [M]. //RAND G. M. Fundamental of Aquatic Toxicology-Effects, Environmental Fate and Risk Assessment [M]. Washington: 2nd ed. R Taylor and Francis.

DOYOTTE A, COSSU C, JACQUIN MC, et al, 1997. Antioxidant enzymes, glutathione and lipid peroxidation as relevant biomarkers of experimental or field exposure in the gills and the digestive gland of the freshwater bivalve Unio tumidus [J]. Aquat. Toxicol. 39: 93-110.

FERNáNDEZ B, CAMPILLO JAC, MARTÍNEZ-GÓMEZ J, 2010. Benedicto Antioxidant responses in gills of mussel (*Mytilus galloprovincialis*) as biomarkers of environmental stress along the Spanish Mediterranean coast [J]. Aquat. Toxicol, 99: 186-197.

FIRAT O, COGUN HY, ASLANYAVRUSU S, et al, 2009. Antioxidant responses and metal accumulation in tissues of Nile tilapia *Oreochromis niloticus* under Zn, Cd and Zn+Cd exposures. J [J]. Appl. Toxicol, 29: 295-301.

FOULKES EC, 1993. Metallothionein and glutathione as determinants of cellular retention and extrusion of cadmium and mercury [J]. Life Sci, 52: 1617-1620.

GERET F, SERAFIM A, BARREIRA L, et al, 2002. Effect of cadmium on antioxidant enzyme activities and lipid peroxidation in the gills of the clam *Ruditapes decussatus* [J]. Biomarkersb, 7: 242-256.

GILL TS, BIANCHI CP, EPPLE A, 1992. Trace metal (Cu and Zn) adaptation of organ systems of the American eel, *Anguilla rostrata*, to external concentrations of cadmium [J]. Comp. Biochem. Physiol. C, 102: 361-371.

JAYAKUMAR P, PAUL VI, 2006. Patterns of cadmium accumulation in selected tissues of the catfish Clarias batrachus (Linn.) exposed to sublethal concentration of cadmium chloride [J]. Vet. Archiv, 76: 167-177.

LEMAIRON P, MATTEWS A, FORLIN L, et al, 1994. Stimulation

ofoxyradical production ofhepaticmicrosomes of flounder (Platichthysfle-sus) and perch (Perca fluviatilis) bymodeland pollutantxenobi-otics [J]. Arch Environ. ContourToxicoh, 26: 191-200.

LIVINGSTONE DR, LIPS F, MARTINEZ PG, et al, 1992. Antioxidant enzymes in the digestive gland of the common mussel (*Mytilus edulis*) [J]. Mar. Biol, 112: 265-276.

MCGEER JC, SZEBEDINSZKY C, MCDONALD DG, et al, 2000. Effects of chronic sublethal exposure to waterborne Cu, Cd or Zn in rainbow trout. 1: Iono-regulatory disturbance and metabolic costs [J]. Aquat. Toxicol, 50: 231-243.

MEISTER A, ANDERSON ME, GLUTATHIONE A, 1983. The Effects of Heavy Metals (other than Mercury) on Marine and Estuarine Organisms [J]. P. Roy. Soc. B-Biol. Sci, 177: 389-410.

MOREIRA SM, GUILHERMINO L, 2005. The use of *Mytilus galloprovincialis* acetylcholinesterase and glutathione S-transferases activities as biomarkers of environmental contamination along the Northwest Portuguese Coast [J]. Environ. Monit. Assess, 105: 309-325.

OCHI T, ISHIGURO T, OHSAWA M, 1983. Participation of active oxygen species in the induction of DNA single-strand scissions by cadmium chloride in cultured Chinese hamster cells [J]. Mut. Res, 122 (2): 169-175.

OHKAWA H, OHISH IN, YAGI K, 1979. Assay for lipid peroxidation in animal tissues by thiobarbituric acid reaction [J]. Anal. Biochem. 95: 351-363.

PANDEY S, PARVEZ S, SAYEED I, et al, 2003. Biomarkers of oxidative stress: a comparative study of river Yamuna fish Wallago attu (Bl. & Schn.) [J]. Sci. Total. Environ, 309: 105-115.

PELGROM SMG, LAMERS LPM, LOCK RAC, et al, 1995. Interactions between copper and cadmium modify metal organ distribution in mature tilapia, *Oreochromis mossambicus* [J]. Environ. Pollut, 90 (3): 415-423.

POWER EA, CHAPMAN PM, 1992. Assessing sediment quality [M]. // BURTON G A. Sediment toxicity assessment. Boca Raton: Lewis Pub.

REGOLI F, HUMMEL H, AMIARD-TRIQUET C, et al, 1998. Trace Metals and Variations of Antioxidant Enzymes in Arctic Bivalve Populations [J]. Arch. Environ. Con. Tox, 35 (4): 594-601.

REMÉO D, BENNANI N, GNASSIA-BARELLI M, et al, 2000. Cadmium and copper display different responses towards oxidative stress in the kidney of the sea bass *Dicentrarchus labrax* [J]. Aquat. Toxicol. 48: 185-194.

ROCHER B, LEGOFF J, PELUHET L, 2006. Genotoxicant accumulation and cellular defence activation in bivalves chronically exposed to waterborne contaminants from the Seine River [J]. Aquat. Toxicol, 79: 65-77.

SAMUEL OA, GEORGE EE, MABEL EI, 2008. Bioaccumulation of cadmium and its biochemical effect on selected tissues of the catfish (*Clarias gariepinus*) [J]. Fish Physiol. Biochem, 34: 61-69.

SEGNER H, BRAUNBECK T, 1988. Hepatocellular adaptation to extreme nutritional conditions in ide, Leuciscus idus melanotus L. (Cyprinidae). A morphofunctional analysis [J]. Fish Physiol. Biochem, 5 (2): 79-97.

SORENSEN EMB, 1991. Metal poisoning in fish. Boca Raton: CRC Press.

TAMÁS L, VALENTOVICOVÁ K, HALUSKOVÁ L, et al, 2009. Effect of cadmium on the distribution of hydroxyl radical, superoxide and hydrogen peroxide in barley root tip [J]. Protoplasma, 236: 67-72.

THOMAS P, WOFFORD HW, 1993. Effects of cadmium and Aroclor 1254 on lipid peroxidation, glutathione peroxidase activity, and selected antioxidants in Atlantic croaker tissues [J]. Aquat. Toxicol, 27: 159-178.

THOMAS PT, WOFFORD HW, NEFF JM, 1982. Effect of cadmium on glutathione content of mullet (*Mugil cephalus*) tissues. In: Vernberg WB, Calabrese A, Thurberg FP, Vernberg FP (eds) Physiological Mechanisms of Marine Pollutant Toxicity [M]. New York: Academic Press.

TORT L, KARGACIN B, TORRES P, et al, 1996. The effect of cadmium exposure and stress on plasm cortisol, metallothionein levels and oxidative status in rainbow trout (*Oncorhynchus mykiss*) liver [J]. Comp. Biochem. Physiol, 114C: 29-34.

USHA RA, RAMAMURTHI R, 1989. Histopathological alterations in the liver of freshwater teleost Tilapia mossambica in response to cadmium toxicity [J]. Ecotoxicol. Environ. Saf, 17: 221-226.

VIARENGO A, 1990. Heavy metal effects on lipid peroxidation in the tissues of *Mytilus galloprovincialis* Lam [J]. Comp. Biochem. Physiol. C, 97: 37-42.

VIARENGO A, 1989. Heavy metals in marine invertebrates, mechanisms of regulation and toxicity at the cellular level [J]. Rev. Aquat. Sci, 1: 295-317.

WARNER HR, 1994. Superoxide dismutase, aging and degenerative disease [J]. Free Radic. Biol. Med, 17: 249-258.

XU HZ, ZHOU CG, MA YA, et al, 2000. Environmental quality of deposits in offshore zone of China [J]. Environ. Protect. Transportation, 21: 16-18.

XU XD, LIN ZH, LI SQ, 2005. The studied of the heavy metal pollution of Jiaozhou Bay Mar [J]. Sci. 29: 48-53.

YU BP, 1994. Cellular defenses against damage from reactive oxygen species [J]. Physiol. Rev, 74: 139-162.

第六章
饲料硒解除镉对皱纹盘鲍毒性作用的研究

第一节 引　　言

目前，中国沿海地区的重金属污染越来越严重，尤其是在沉积物中，具有致癌作用的 Cd 在很多的海湾如锦州湾、渤海湾、胶州湾都超出了国家的水质标准。Cd 本身不能直接导致 ROS 的产生（Nemmiche 等，2007），但是据报道 Cd 可以间接诱导 ROS 的产生（Galán 等，2001；Tamás 等，2009），由 Cd 导致的氧化应激影响 DNA、RNA、核糖体的合成，由此而导致一些酶的失活（Stohs 等，2000）。Cd 对鱼、贝类、甲壳类等均可产生氧化应激，对其产生不利的影响。ROS 的大量产生会导致机体的抗氧化系统发挥作用，抗氧化系统主要包括小分子的抗氧化物质（维生素 E、维生素 C、Se、GSH）和抗氧化酶（CAT、SOD、GPx、GR、GST），抗氧化系统和活性氧之间的动态平衡是保证机体健康状态的重要因素，如果这种平衡向氧化一边倾斜，就会出现氧化应激，对生物体产生危害。Cd 穿过细胞膜到达细胞质，导致一系列的 ROS 的产生，从而导致脂质的过氧化。Gagné 等（2008）报道淡水贻贝在 Cd 的处理下，鳃组织和外套膜组织的脂质过氧化物（LPO）及 DNA 发生断裂。抗氧化系统能够消除一些 ROS 从而达到解毒的目的。众所周知，贝类对金属离子的积

第六章 饲料硒解除镉对皱纹盘鲍毒性作用的研究

累能力远远大于其他的水生生物。既然重金属 Cd 可以对水产动物产生氧化应激，使其产生氧化损伤，则可以通过添加抗氧化的物质，提高水产动物的抗氧化能力来抵抗重金属的氧化胁迫。

通过笔者实验室前期的研究发现，饲料中添加抗氧化的物质（维生素 A、维生素 E、维生素 D、硒、GSH、硫辛酸、Fe、Zn、Cu），均提高了皱纹盘鲍抗氧化水平，但是添加抗氧物质能否抵抗外界环境因子的氧化应激还有待进一步研究。

硒是动植物必需的微量元素，是潜在的抗氧化剂，是体内抗氧化体系的组成部分，常作为 GPx 的活性中心参与抗氧化作用（Wang 等，2007）。在体内，大部分硒以硒半胱氨酸的形式与多种重要蛋白结合，通过硒蛋白行使各种生理功能，其中研究较多的是其免疫和抗氧化作用。硒的抗氧化作用和硒蛋白有关，硒蛋白主要包括谷胱甘肽过氧化物酶、硒蛋白 P、硫氧还蛋白还原酶，适量的硒能改善体内的抗氧化系统，促进抗氧化水平的提高（王蔚芳，2009）。在小鼠上研究发现硒可以降低 Cd 对肝脏（Ognjanovic 等，2007；Newairy 等，2007；El Heni Jihen 等，2009；Jamba 等，2000）、肾脏（El-Sharaky 等，2007；Ognjanovic 等，2007）造成的抗氧化系统损伤。Jihen（2008）研究发现，硒可以部分降低小鼠肝脏 Cd 积累及缓解 Cd 对肝脏造成的组织损伤。但是硒缓解 Cd 毒性的具体机制并不清楚。

从 Cd 毒性产生的机制来看，为了抵抗重金属 Cd 的毒性，机体可通过如下的途径来解 Cd 毒：①Cd 和金属硫蛋白具有很强的亲和性，金属硫蛋白和 Cd 结合可以降低 Cd 的毒性。②Se 可以诱导金属硫蛋白的合成。③具有抗氧化活性的物质可以通过降低 ROS 的含量来降低 Cd 的毒性。本实验以皱纹盘鲍为研究对象，主要研究硒在 Cd 解毒中的作用及探索它们解毒的机制，这样既可以加强鲍的养殖管理，又可以提高水产动物的食物安全水平。

第二节 摄食生长实验设计

(一) 实验动物和养殖管理设计

皱纹盘鲍幼鲍购买于青岛鳌山卫育苗场,为同一批鲍苗。在中国海洋大学水产馆养殖系统中暂养 2 周后挑选大小一致的健康个体[平均体重为 (3.17±0.01) g] 随机分成 3 组,每组 3 个重复,每个重复 50 只鲍。生长实验在中国海洋大学水产馆养殖系统中进行,静水养殖 2 个月。实验期间,每天换水两次,每次换水量为实验缸水量的一半。每天 17:00 投喂人工饲料,次日 8:00 清底,并密切观察采食及健康状况。养殖过程中水温 18~21℃,盐度为 22~28,pH 为 7.4~7.9,溶氧量大于 6mg/L,水体中硒含量为 0.46μg/L。

(二) 生长实验设计

根据前面的实验,笔者团队得知重金属 Cd 对皱纹盘鲍的半致死浓度为 2.848mg/L。笔者团队选择半致死浓度的 1/8 作为本实验的浓度,即选择 0.35mg/L 作为本实验 Cd 的浓度。Cd 的实测值为 (0.34±0.01) mg/L,Cd 的添加形式为 $CdCl_2 \cdot 2.5H_2O$。饲料中添加硒的水平为 0、1.5、4.5mg/kg,硒源为 $Na_2SeO_3 \cdot 5H_2O$。各饲料组中硒的实测值分别为:0.1、0.95、4.2mg/kg。基础饲料配方参照 Mai 等 (2003),为半精制饲料,蛋白源为明胶和酪蛋白,脂肪源为鲱鱼油和大豆油,两者的比例为 1:1,主要糖源为糊精,再辅以纤维素、维生素和矿物质(不含硒)等配制而成。基础饲料中的常规指标,包括粗蛋白质、粗脂肪和粗灰分的测定参照 AOAC (1995)。皱纹盘鲍的饲料配方及其营养成分见表 6-1。皱纹盘鲍饲料的具体配制步骤和保存方法参考 Zhang 等 (2007)。

(三) 取样与样品处理

养殖实验结束时,皱纹盘鲍禁食 3d,排空其肠道内容物。实验缸中所有的皱纹盘鲍称重计数后,收集肝胰脏、肌肉、鳃、外套膜、贝壳。肝胰脏、肌肉、鳃和外套膜剪成小块混匀后分装在小管中,保存在 -80℃ 冰箱中待测。离心得到的血清立即放在液氮中速

第六章 饲料硒解除镉对皱纹盘鲍毒性作用的研究

表 6-1 基础饲料配方及其营养成分

成分	含量（%）
酪蛋白[a]	25.00
明胶[b]	6.00
糊精[b]	33.50
羧甲基纤维素[b]	5.00
海藻酸钠[b]	20.00
维生素混合物[c]	2.00
矿物质混合物[d]	4.50
氯化胆碱[b]	0.50
豆油：鲱鱼油[e]	3.50
概略养分分析（以干重计）	
粗蛋白质	29.41
粗脂肪	3.26
粗灰分	10.01

注：a. sigma 公司。

b. 国药集团上海化学试剂有限公司。

c. 每 1 000g 饲料中含有：盐酸硫胺素，120mg；核黄素，100mg；叶酸，30mg；盐酸吡哆素，40mg；烟酸，800mg；泛酸钙，200mg；肌醇，4 000mg；生物素，12mg；维生素 B_{12}，0.18mg；维生素 C，4 000mg；维生素 E，450mg；维生素 K_3，80mg；维生素 A，100 000IU；维生素 D，2 000IU。

d. 每 1 000g 饲料中含有：NaCl，0.4g；$MgSO_4 \cdot 7H_2O$，6.0g；$NaH_2PO_4 \cdot 2H_2O$，10.0g；KH_2PO_4，20.0g；$Ca(H_2PO_4)_2 \cdot H_2O$，8.0g；Fe-柠檬酸，1.0g；$ZnSO_4 \cdot 7H_2O$，141.2mg；$MnSO_4 \cdot H_2O$，64.8mg；$CuSO_4 \cdot 5H_2O$，12.4mg；$CoCl_2 \cdot 6H_2O$，0.4mg；KIO_3，1.2mg。

e. 豆油：鲱鱼油＝1∶1。

冻后放－80℃冰箱中保存。肝胰脏立即放入无 RNA 酶管中，液氮速冻后放入－80℃冰箱。

肝胰脏样品使用前解冻，加入预冷的 0.86% 生理盐水中，冰上匀浆，然后 4℃、4 000r/min 离心 15min，取肝胰脏上清测定抗氧化指标。以特定增长率（SGR）衡量鲍的生长情况：

$$SGR = 100 \times (\ln W_t - \ln W_i)/t$$

式中，W_t、W_i 分别代表鲍的终末体重和初始体重（g），t 代表时间（d）。

第三节　饲料硒对镉胁迫下皱纹盘鲍生长和组织镉含量的影响

（一）饲料中硒的溶失结果

含硒 1.5mg/kg 饲料组 [实测 (1.30±0.13) mg/kg] 经过 3、6h 的溶失后的硒含量分别为 (1.04±0.12)、(0.95±0.05) mg/kg，均没有显著性差异。含硒 4.5mg/kg 饲料组 [实测 (4.17±0.15) mg/kg] 经过 3、6h 的溶失后的硒含量分别为 (3.99±0.29)、(3.88±0.12) mg/kg，均没有显著性差异。

（二）饲料中的硒对 Cd 胁迫下皱纹盘鲍生长的影响

如表 6-2 所示，饲料中不同的 Se 添加水平对 Cd 胁迫下皱纹盘鲍生长和存活率无显著性影响（$P>0.05$）。各组皱纹盘鲍的存活率为 90.83%～93.33%；各组生长虽然没有统计学的显著性差异，但是 4.5mg/kg Se 饲料组的 SGR 最大。

表 6-2　饲料中的硒（Se）对 Cd 胁迫下的皱纹盘鲍生长和存活的影响（平均值±标准误，$n=3$）

饲料中硒含量 (mg/kg)	初始体重 (g)	终末体重 (g)	特定增长率 (%)	存活率 (%)
对照	3.17±0.01	3.88±0.09	0.34±0.04	92.50±1.44
1.5	3.17±0.01	4.01±0.14	0.39±0.05	90.83±3.01
4.5	3.15±0.01	3.83±0.10	0.33±0.04	93.33±2.21
One-way ANOVA				
P 值	0.425	0.690	0.563	0.748
F 值	0.989	0.537	0.632	0.304

注：同列数据栏中，经 Turkey 检验差异不显著的平均值之间用相同的字母表示（$P>0.05$）。

（三）饲料中的硒对镉胁迫下皱纹盘鲍肝胰脏组织镉含量的影响

如表 6-3 所示，饲料中的硒显著性影响皱纹盘鲍肝胰脏、肌

第六章 饲料硒解除镉对皱纹盘鲍毒性作用的研究

表6-3 饲料中的硒（Se）对Cd胁迫下的皱纹盘鲍组织Cd含量的影响（平均值±标准误，$n=3$）

饲料中硒含量 (mg/kg)	血清 (μg/mL)	贝壳 (μg/g)	肌肉 (μg/g)	外套膜 (μg/g)	鳃 (μg/g)	肝胰脏 (μg/g)
对照	1.94±0.08[a]	19.49±3.69	17.34±17.34[a]	37.32±0.46[a]	69.66±3.57[a]	418.84±21.31[a]
1.5	1.33±0.10[b]	16.39±1.92	14.85±14.85[ab]	33.19±0.02[b]	56.42±2.15[b]	306.85±9.44[b]
4.5	1.38±0.12[b]	21.14±3.35	11.35±11.35[b]	32.66±0.65[b]	56.55±2.57[b]	253.07±19.40[b]
One-way ANOVA						
P值	0.009	0.573	0.012	0.00*	0.025	0.000
F值	11.528	0.611	10.063	30.922	7.246	41.251

注：同列数据栏中，经Turkey检验差异不显著的平均值之间用相同的字母表示（$P>0.05$）。

肉、血清、鳃和外套膜中的 Cd 含量。在血清、肝胰脏、外套膜和鳃中，1.5、4.5mg/kg 处理组的 Cd 含量显著低于对照组。在肌肉组织中，4.5mg/kg 处理组的 Cd 含量显著低于 1.5mg/kg 硒的处理组及对照组，但是 1.5mg/kg 硒饲料组和对照组无显著性差异。贝壳中 Cd 的含量在各实验组均无显著性差异。

第四节　饲料硒对镉胁迫下皱纹盘鲍肝胰脏抗氧化指标的影响

（一）饲料硒对镉胁迫下皱纹盘鲍肝胰脏抗氧化指标的影响

如表 6-4 所示，各处理组的肝胰脏 SOD、CAT、GST 的活力和 Trx 的含量无显著性差异。但是 1.5、4.5mg/kg 实验组肝胰脏 CAT 和 GST 的活力高于对照组。

饲料中的 Se 显著影响了肝胰脏 Se-GPx、TrxR、TrxP 的活力及 GSH 的含量。1.5、4.5mg/kg 饲料组的肝胰脏 Se-GPx、TrxR、TrxP 的活力和 GSH 含量均显著高于对照组，但是 1.5、4.5mg/kg 处理组之间无显著性差异。

最近几年来，硒与镉相互作用的研究已有了很大的进展，适量的硒对镉引起的毒害起拮抗作用（Newairy 等，2007；Tran 等，2007）。但硒拮抗镉中毒的机制还不清楚，现在公认的学说是自由基学说。Cd 本身虽然不能直接导致 ROS 的产生（Nemmiche 等，2007），但是据报道 Cd 可以间接诱导 ROS 的产生（Galan 等，2001；Tamás 等，2009），由 Cd 导致的过量的 ROS 影响到 DNA、RNA、核糖体的合成，由此而导致一些酶的失活（Stohs 等，2000）。硒能够拮抗 Cd 的一个原因是：硒是天然的抗氧化剂，是 GSH-Px 的活性中心（Rotruck 等，1973），其主要功能就是它的抗氧化性，硒可以提高机体的抗氧化能力来抵抗 Cd 中毒过程中产生的活性氧。根据李宣海等（1999）报道，给大鼠补充适量的硒（0.2mg/kg）在提高 GPx 活性的同时，可以降低血清中的脂质过氧化水平。Michiels 等（1994）报道，硒在消除 MDA 的过程中同超氧化物歧化酶（SOD）

表 6-4 饲料中的硒 (Se) 对 Cd 胁迫下的皱纹盘鲍肝胰脏抗氧化相关指标的影响（平均值±标准误，$n=3$）

饲料中硒含量 (mg/kg)	SOD	CAT	Se-GPx	GST	GSH	Trx	TrxR	TrxP
0	10.64±0.75	3.25±0.35	4.76±0.27[b]	2.53±0.03	5.88±0.14[b]	292.64±0.57	2.51±0.03[b]	5.38±0.16[b]
1.5	10.30±0.50	4.18±0.70	9.03±1.12[a]	3.05±0.23	7.64±0.06[a]	291.95±1.07	2.97±0.02[a]	6.61±0.28[a]
4.5	10.47±0.42	4.54±0.08	8.00±0.38[a]	2.90±0.05	7.21±0.11[a]	290.67±1.18	2.84±0.07[a]	7.17±0.30[a]
One-way ANOVA								
P 值	0.917	0.198	0.012	0.089	0.000	0.053	0.001	0.007
F 值	0.088	2.144	10.174	3.730	71.627	4.992	28.729	12.840

注：同列数据栏中，经 Turkey 检验差异不显著的平均值之间用相同的字母表示（$P>0.05$）。

和过氧化氢酶（CAT）有协同作用。贺宝芝等（1998）发现，添加硒可以显著提高铅中毒的大鼠红细胞 SOD 的活性，从而抑制了脂质过氧化水平，并且对铅在组织中的积累具有一定的抑制作用。马卓等研究表明，添加硒对绵羊的 Pb、Cd 联合中毒具有显著的拮抗作用，表现在添加硒组各器官组织的病理损伤减轻。在本研究中，笔者团队对 Cd 胁迫下的硒抗氧化机制进行了全面的研究，包括一些抗氧化酶（SOD、CAT、Se-GPx、GST）和抗氧化小分子（GSH），还有硫氧还蛋白系统的三个成分——TrxR、TrxP 和 Trx。在实验中，Se 的添加显著提高了 Cd 胁迫下肝胰脏的两种含硒酶（Se-GPx 和 TrxR）的活力，相似的研究表明硒的处理可以升高哺乳动物、鱼类和水生无脊椎动物的含硒蛋白 Se-GPx 或 TrxR 的活力（Zhang 等，2008；Adesiyan 等，2011；Atencio 等，2009；Tran 等，2007；Wang 等，2010）。本实验研究发现，Se 的添加显著提高了肝胰脏 GSH 的含量。GSH 被认为是细胞防御中最重要的抗氧化剂（Meister 和 Anderson，1983；Segner 和 Braunbeck，1998）。这个富含巯基的三肽物质通过与金属离子螯合或改变金属的吸收来达到缓解金属毒性的作用（Foulkes，1993；Ochi 等，1988；Burton 等，1995）。抗氧化小分子 GSH 可将亚硒酸盐还原为硒化合物，后者有高度的亲脂性从而改变了镉在关键组织中的分布与毒性。总之，硒的添加可以增强 Cd 胁迫下皱纹盘鲍肝胰脏的抗氧化能力。

（二）饲料硒对镉胁迫下皱纹盘鲍肝胰脏抗氧化指标的影响

如表 6-5 所示，含硒 1.5、4.5mg/kg 饲料组的肝胰脏 MDA 的含量和蛋白羰基含量均显著低于对照组，但是 1.5、4.5mg/kg 饲料组之间无显著性差异。1.5、4.5mg/kg 组 F 值均高于对照组。F 值越大，DNA 的断裂程度越小。因此，1.5、4.5mg/kg 组的 DNA 断裂程度都低于对照组，且 1.5mg/kg 饲料组显著低于 4.5mg/kg 组。

Cd 的胁迫可以导致生物体的过量 ROS 的产生，这些 ROS 如果不及时清除，就会对生物体的大分子物质如脂肪、蛋白质甚至是

DNA 产生损害（Stohs 等，2000）。很多研究表明，Cd 污染会导致软体动物脂质过氧化水平的升高（Geret 等，2002b；Company 等，2004；Chandran 等，2005）。在 5mg/L Cd 离子胁迫 33d 后，大西洋黄鱼（*Micropogonias undulates*）肝脏组织中脂质过氧化的水平升高（Thomas 和 Wofford，1993）。过量的 ROS 可以导致一些非酶蛋白的变性，包括蛋白糖基化水平，被广泛应用为氧化应激的标志物（Shacter 等，1994；Je 等，2008）。根据 Kaloyianni 等（2009）研究表明，在不同浓度的金属离子下，贻贝 *M. galloprovincialis* 红细胞中的蛋白羰基含量升高。Banaflvi 等研究了 Cd 与 DNA 修复的关系，发现 Cd 可以引起 DNA 单链的断裂和 8-羟基脱氧鸟苷的生成。在本实验中，含硒 1.5、4.5mg/kg 饲料组的肝胰脏 MDA 和蛋白羰基含量均显著低于对照组。1.5mg/kg 饲料组的 DNA 的断裂水平要显著高于对照组。根据 Tran 等（2007）的研究表明，$4\mu g/L$ 的硒可以降低重金属导致的 DNA 的损伤。体外实验均表明，$5.0\mu mol/kg$ 的硒可抑制镉所诱导的 DNA 损伤（余日安等，2003；余日安和陈雪敏，2004）。这些研究都表明，硒可以降低机体因 Cd 胁迫导致的氧化应激水平，在一定程度上降低了 Cd 对生物体的损害。

表 6-5 饲料中的硒（Se）对 Cd 胁迫下皱纹盘鲍肝胰脏氧化指标的影响（平均值±标准误，$n=3$）

饲料中硒含量 (mg/kg)	MDA (nmol, 每毫克蛋白中)	蛋白质羰基化 (nmo, 每毫克蛋白中)	F 值
对照	24.11 ± 0.29^a	8.67 ± 0.23^a	0.32 ± 0.05^b
1.5	22.07 ± 0.10^b	4.35 ± 0.32^b	0.70 ± 0.05^a
4.5	21.63 ± 0.53^b	5.79 ± 0.43^b	0.49 ± 0.09^{ab}
One-way ANOVA			
P 值	0.005	0.000	0.032
F 值	14.046	42.939	6.449

注：同列数据栏中，经 Turkey 检验差异不显著的平均值之间用相同的字母表示（$P>0.05$）。

第五节 饲料硒对镉胁迫下皱纹盘鲍肝胰脏金属硫蛋白（MT）含量和 MTF-1 mRNA 表达的影响

如图 6-1 所示，饲料中的 Se 显著了影响皱纹盘鲍肝胰脏 MT 的含量，含硒 1.5、4.5mg/kg 的饲料组皱纹盘鲍肝胰脏 MT 的含量均显著高于对照组，且随着 Se 含量的增加呈上升的趋势，但 1.5、4.5mg/kg 的饲料组之间无显著性差异。1.5、4.5mg/kg 饲料组中的肝胰脏 MT 和 MTF-1 的 mRNA 的表达量均高于对照组，但与对照组无显著性的差异。

金属硫蛋白是一类广泛存在于机体内、富含半胱氨酸、与金属离子具有高度亲和性的低分子质量蛋白，MT 与 Cd 等有毒重金属离子结合能减少其对组织的损害，具有解毒、调节机体抗氧化功能的作用，是影响 Cd 毒性作用的重要因素之一（Dabrio 等，2002）。研究证明，MT 具有优于其他抗氧化剂的保护作用（赵红光和龚守良，2005）。Se 是机体合成 MT 的有效诱导物（田晓丽等，2006；郭家彬等，2006），很多学者认为硒对 Cd 毒性作用有一定拮抗作用的原因在于 Se 能够诱导 MT 合成。本实验研究发现，1.5mg/kg 和 4.5mg/kg 饲料组的肝胰脏 MT 含量显著高于对照组，且和肝胰脏 Cd 含量的变化趋势一致。以此可以说明在硒拮抗重金属 Cd 毒性中，金属硫蛋白发挥了重要的作用。

研究表明，贝类更容易积累重金属，并且 Cd 大量积累可以导致肝脏细胞的坏死。研究表明，Se 通过降低组织中 Cd 的含量来保护 Cd 对机体的危害（Chen 等，1975；Jamba 等，2000）。但是根据 Meyer 等（1982）和 Stajn 等（1997）的研究发现，硒没有降低组织中 Cd 的含量，他们认为能否降低组织 Cd 的含量取决于 Se/Cd 的比例。在本实验中，饲料中的硒可以显著降低皱纹盘鲍肝胰脏、血清、鳃、肌肉和外套膜中的 Cd 含量。王宗元等（1997）通过 ^{109}Cd 标记的 $CdCl_2$ 进行的体外实验认为，硒拮抗 Cd 中毒的机制

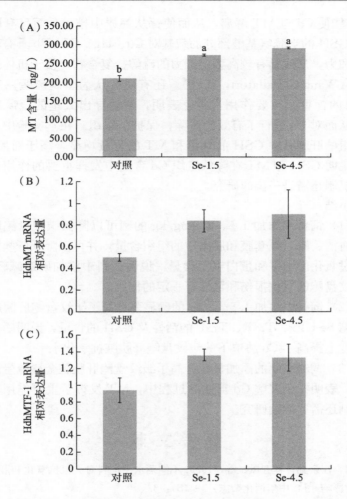

图 6-1 不同浓度下 Se 对皱纹盘鲍镉胁迫下的肝胰腺中 MT 含量、Hdh-MT 和 HdhMTF-1 基因的表达的影响（平均值±标准误，$n=3$）

（A）不同浓度下的 Se 对 Cd 胁迫下皱纹盘鲍肝胰脏 MT 含量的影响；（B）不同浓度下的 Se 对 Cd 胁迫下皱纹盘鲍肝胰脏 MT mRNA 表达的影响；（C）不同浓度下的 Se 对 Cd 胁迫下皱纹盘鲍肝胰脏 MTF-1 mRNA 表达的影响。不同小写字母表示差异显著（$P<0.05$），相同或无字母表示差异不显著（$P>0.05$）。

为 Se 能使 Cd 与 MT 解离，从而使镉从粪尿中排出。而且有研究报道 GSH 的半胱氨基酸部分的巯基对 Cu、Hg、Cd、Pb 具有较高的亲和力，形成具有较高稳定能力的硫醇盐复合物，便于机体排出体外（Wang 和 Ballatori，1998）。还有研究认为金属硫蛋白对降低 Cd 的含量起重要作用，研究表明，硒诱导出大量的金属硫蛋白，从而对 Cd 进行了有效的清除，保护了组织。在本实验中，硒添加组的肝胰脏中 GSH 的含量和 MT 的含量均显著高于对照组，可以说明 GSH 和 MT 在降低组织 Cd 含量中发挥重要的作用，具体的机制还需进一步的研究。

小结：

（1）饲料中添加 1.5、4.5mg/kg 的硒可以降低皱纹盘鲍肝胰脏、血清、鳃、外套膜和肌肉中的组织含量，并可以降低肝胰脏中脂质过氧化的水平和蛋白羰基含量，说明饲料中添加硒对减轻 Cd 对皱纹盘鲍的氧化损伤和毒性有一定的作用。

（2）饲料中添加 1.5mg/kg 的硒显著升高了皱纹盘鲍肝胰脏抗氧化酶 Se-GPx、TrxR、TrxP 的活性及 GSH 的含量，说明硒在一定程度上提高了 Cd 胁迫下的皱纹盘鲍肝胰脏抗氧化水平。

（3）饲料中硒的添加显著升高了皱纹盘鲍肝胰脏金属硫蛋白的含量，表明在硒缓解 Cd 毒性的过程中，MT 发挥了重要作用，具体机制还需要继续研究。

参 考 文 献

李宣海，汪余勤，程五凤，等，1999. 不同硒水平饲料对大鼠抗氧化和肝纤维化的影响 [J]. 中华消化杂志，19（5）：3.

郭军华，徐卓立，吴德政，1993. 诱导生物合成金属硫蛋白减轻顺铂对小鼠的致死毒性 [J]. 癌症，12（6）：476.

王蔚芳，麦康森，张文兵，等，2007. 皱纹盘鲍（*Haliotis discus hannai* Ino）对饲料中硒的需要量及其免疫反应的研究 [C] //. 2007 年中国水产学会学术年会暨水产微生态调控技术论坛论文摘要汇编：66-67.

余日安，陈学敏，2003. 低剂量的硒与镉联合作用对大鼠肝细胞 DNA 损伤的影响 [J]. 毒理学杂志，17（4）：200-202.

第六章 饲料硒解除镉对皱纹盘鲍毒性作用的研究

余日安, 陈学敏, 2004. 硒对镉诱导的在体大鼠肝细胞 DNA 损伤、细胞凋亡和细胞增殖的影响 [J]. 中华预防医学杂志, 38 (3): 4.

周杰昊, 程时, 1995. 金属硫蛋白与医学 [J]. 生理科学进展, 26 (1): 29-34.

ADESIYAN AC, OYEJOLA TO, ABARIKWU SO, et al, 2011. Selenium provides protection to the liver but not the reproductive organs in an atrazine-model of experimental toxicity [J]. Exp. Toxicol. Pathol, 63: 201-207.

ATENCIO L, MORENO I, JOS A, et al, 2009. Effects of dietary selenium on the oxidative stress and pathological changes in tilapia (*Oreochromis niloticus*) exposed to a microcystin-producing cyanobacterial water bloom [J]. Toxicon, 53: 269-282.

BERGGRE MM, MANGIN JF, GASDAKA JR, et al, 1999. Effect of selenium on rat thioredoxin reductase activity: increase by supranutritional selenium and decrease by selenium deficiency [J]. Biochem. Pharmacol, 57 (2): 187-193.

CERKLEWSKI FL, FORBES RM, 1976. Influence of dietary zinc on lead toxicity in rats [J]. J. Nutr, 106: 689-696.

CHELOMIN VP, BELCHEVA NN, 1991. Alterations of microsomal lipid synthesis in gill cells of bivalve mollusk *Mizuhopecten yessoensis* in response to cadmium accumulation [J]. Comp. Biochem. Physiol (C), 99: 1-5.

CHEN R W, WHANGER P D, WESWIG P H, 1975. Selenium-induced redistribution of cadmium binding to tissue proteins: a possible mechanism of protection against cadmium toxicity [J]. Bioinorganic Chemistry, 4 (2): 125-133.

COMBS GF, COMBS SB, 1984. The nutritional biochemistry of selenium [J]. Ann. Rev. Nutr, 4: 257-280.

DABRIO M, AR RODRÍGUEZ, 2002. Characterization of Zinc Metallothioneins Using Electroanalytical Methods [J]. Trace Elements in Man and Animals, 10: 1116.

DE ZOYSA M, LEE J, 2007. Two ferritin subunits from disk abalone (*Haliotis discus discus*): Cloning, characterization and expression analysis [J]. Fish Shellfish Immun, 23 (3): 624-635.

DE ZOYSA M, WHANG I, LEE Y, et al, 2009. Transcriptional analysis of antioxidant and immune defense genes in disk abalone (*Haliotis discus discus*) during thermal, low-salinity and hypoxic stress [J]. Comp. Biochem.

Physiol. (B), 154: 387-395.

FILIPI M, HEI T K, 2004. Mutagenicity of cadmium in mammalian cells: implication of oxidative DNA damage [J]. Mutat. Res/fund Mol. M, 546 (1-2): 81-91.

FINKEL T, HOLBROOK NJ, 2000. Oxidants, oxidative stress and the biology of ageing [J]. Nature, 408: 239-247.

GABBIANELLI R, LUPIDI G, VILLARINI M, et al, 2003. DNA damage induced by copper on erythrocytes of gilthead sea bream *Sparus aurata* and mollusk *Scapharca inaequivalvis* [J]. Arch. Environ. Contam. Toxicol, 45: 350-356.

GAGNÉ A F, AUCLAIR A J, TURCOTTE A P, et al, 2008. Ecotoxicity of CdTe quantum dots to freshwater mussels: Impacts on immune system, oxidative stress and genotoxicity-ScienceDirect [J]. Aquat. Toxicol, 86 (3): 333-340.

GALÁN A, TROYANO A, VILABOA N E, et al, 2001. Modulation of the stress response during apoptosis and necrosis induction in cadmium-treated U-937 human promonocytic cells [J]. Biochimica et Biophysica Acta (BBA) - Molecular Cell Research, 1538 (1): 38-46.

GERET F, SERAFIM A, BARREIRA L, et al, 2002b. Effect of cadmium on antioxidant enzyme activities and lipid peroxidation in the gills of the clam *Ruditapes decussatus* [J]. Biomarkersb, 7: 242-256.

HARRIS ED, 1991. Copper transport: an overview [J]. Exp. Biol. Med, 196 (2): 130-140.

HERMES-LIMA M, WILLMORE WG, STOREY KB, 1995. Quantification of lipid peroxidation in tissue extracts based on Fe (Ⅲ) xylenol orange complex formation [J]. Free Radical Bio. Med, 19: 271-280.

JAMBA L, NEHRU B, BANSAL M P, 2000. Effect of selenium supplementation on the influence of cadmium on glutathione and glutathione peroxidase system in mouse liver [J]. J. Trace Elem. Expe. Med, 13 (3): 299-304.

JE JH, LEE TH, KIM DH, et al, 2008. Mitochondrial ATP synthase is a target for TNBS-induced protein carbonylation in XS-106 dendritic cells [J]. Proteomics, 8: 2384-2393.

JI X, WANG W, CHENG J, et al, 2006. Free radicals and antioxidant status in rat liver after dietary exposure of environmental mercury [J]. Environ. Toxicol. Pharmacol, 22: 309-314.

JIHEN E H, IMED M, FATIMA H, et al, 2009. Protective effects of selenium (Se) and zinc (Zn) on cadmium (Cd) toxicity in the liver of the rat: Effects on the oxidative stress [J]. Ecotox Environ. Safe, 72 (5): 1559-1564.

JIHEN E H, IMED M, FATIMA H, et al, 2008. Protective effects of selenium (Se) and zinc (Zn) on cadmium (Cd) toxicity in the liver and kidney of the rat: Histology and Cd accumulation [J]. Food Chem. Toxicol, 46 (11): 3522-3527.

JIMéNEZ I, SPEISKY H, 2000. Effects of copper ions on the free radicalscavenging properties of reduced glutathione: implications of a complex formation [J]. Trace Elem. Biol. Med, 14: 161-167.

KALOYIANNI M, DAILIANIS S, CHRISIKOPOULOU E, et al, 2009. Oxidative effects of inorganic and organic contaminants on haemolymph of mussels [J]. Comp. Biochem. Physiol. Part C, 149: 631-639.

MAI K, ZHANG W, TAN B, HE G, et al, 2003. Effects of dietary zinc on the shell biomineralization in abalone *Haliotis discus hannai* Ino [J]. J. Exp. Mar. Biol. Ecol, 283: 51-62.

MCPHERSON A, 1994. Selenium vitamin E and biological oxidation [M]. // COLE DJ, GARNSWORTHY PJ. Recent advances in animal nutrition. Oxford: Butterworth and Heinemann's.

MEYER SA, HOUSE WA, WELCH RM, 1982. Some metabolic interrelationships between toxic levels of cadmium and nontoxic levels of selenium fed to rats [J]. J. Nutr, 112 (5): 954-961.

NEMMICHE S, D CHABANE-SARI, GUIRAUD P, 2007. Role of alpha-tocopherol in cadmium-induced oxidative stress in Wistar rat's blood, liver and brain [J]. Chem-Biol Interact, 170 (3): 221-230.

NEWAIRY AA, EL-SHARAKY AS, BADRELDEEN MM, et al, 2007. The hepatoprotective effects of selenium against cadmium toxicity in rats [J]. Toxicology, 242: 23-30.

OCHI T, OTSUKA F, TAKAHASHI K, et al, 1988. Glutathione and metallothioneins as cellular defense against cadmium toxicity in cultured Chinese hamster cells [J]. Chem. Biol. Interact, 65: 1-14.

ORUN I, ATES B, SELAMOGLU Z, et al, 2005. Effects of various sodium selenite concentrations on some biochemicaland hematological parameters of rainbow trout (*Oncorhynchus mykiss*) [J]. Fresen. Environ. Bull, 14: 18-22.

Remacle J, Lambert D, Raes M, et al, 1992. Importance of various antioxidant enzymes for cell stability. Confrontation between theoretical and experimental data. [J]. Pathophysiology, 286: 41-46.

RICHARDS M, 1989. Recent developments in trace element metabolism and function: role of metallthionein in copper and zinc metabolism [J]. J. Nutr, 119: 1062-1070.

ROTRUCK J T, POPE A L, GANTHER H E, et al, 1973. Selenium: biochemical role as a component of glutathione peroxidase [J]. Science, 179 (4073): 588-90.

SAITO Y, TAKAHASHI K, 2002. Characterization of selenoprotein P as a selenium supply protein [J]. Eur. J. Biochem, 269: 5746-5751.

SHACTER E, WILLIAMS JA, LIM M, et al, 1994. Differential susceptibility of plasma proteins to oxidative modification: examination by western blot immunoassay [J]. Free Radical Bio. Med, 17: 429-437.

STAJN A, ZIKIĆ RV, OGNJANOVIĆ B, et al, 1997. Effect of cadmium and selenium on the antioxidant defense system in rat kidneys [J]. Comp. Biochem. Physiol. C Pharmacol. Toxicol. Endocrinol, 117 (2): 167-172.

STOHS SJ, BAGCHI D, HASSOUN E, et al, 2000. Oxidative mechanisms in the toxicity of chromium and cadmium ions [J]. J. Environ. Pathol. Oncol, 19: 201-213.

TALLANDINI L, CECCHI R, DE BONI S, et al, 1996. Toxic levels of selenium in enzymes and selenium uptake in tissues of a marine fish [J]. Biol. Trace Elem. Res, 51: 97-106.

TAMÁS L, VALENTOVICOVÁ K, HALUSKOVÁ L, et al, 2009. Effect of cadmium on the distribution of hydroxyl radical, superoxide and hydrogen peroxide in barley root tip [J]. Protoplasma, 236: 67-72.

THORNALLEY PJ, VASAK M. 1985. Possible role for metallothionein in protection against radiation-induced oxidative stress. Kinetics and mechanism of its reaction with superoxide and hydroxyl radicals [J]. Biochim. Biophys. Acta, 827: 36-44.

TRAN D, MOODY AJ, FISHER AS, 2007. Protective effects of selenium on mercury-induced DNA damage in mussel haemocytes [J]. Aquat. Toxicol, 84: 11-18.

TREVISAN R, MELLO DF, FISHER AS, et al, 2011. Selenium in water enhances antioxidant defenses and protects against copper-induced DNA damage in the blue mussel *Mytilusedulis* [J]. Aquat. Toxicol, 101: 64-71.

TREVISAN R, MELLO DF, FISHER AS, 2011. Selenium in water enhances antioxidant defenses and protects against copper-induced DNA damage in the blue mussel *Mytilusedulis* [J]. Aquat. Toxicol, 101: 64-71.

WANG HW, XU HM, XIAO GH, et al, 2010. Effects of selenium on the antioxidant enzymes response of *Neocaridina heteropoda* exposed to ambient nitrite [J]. Bull. Environ. Contam. Toxicol, 84: 112-117.

WANG W, BALLATORI N, 1998. Endogenous glutathione conjugates: occurrence and biological functions. Pharmacol Rev 50: 335-356 [J]. Pharmacol. Rev, 50 (3): 335-356.

WINTERBOURN CC, 1982. Superoxide-dependent production of hydroxyl radicals in the presence of iron salts (Letter) [J]. Biochem. J, 205: 463.

YUAN X, TANG C, 1999. Lead effect on DNA and albumin in chicken blood and the protection of selenium nutrition [J]. J. Environ. Sci. Health A, 34: 1875-1887.

ZHANG JS, WANG HL, PENG DG, 2008. Further insight into the impact of sodium selenite on selenoenzymes: high-dose selenite enhances hepatic thioredoxin reductase 1 activity as a consequence of liver injury [J]. Toxicol. Lett, 176: 223-229.

ZHANG W, MAI K, XU W, 2007. Interaction between vitamins A and D on growth and metabolic responses of abalone *Haliotis discus hannai*, Ino [J]. J. Shellfish Res, 26: 51-58.

第七章
饲料硫辛酸解除镉对皱纹盘鲍毒性作用的研究

第一节 引 言

α-硫辛酸是广泛存在于动植物的一种强还原物质,被誉为万能抗氧化剂(Packer 等,1995)。硫辛酸能被消化道轻易吸收,从食物中吸收的硫辛酸经转运到达机体的各个组织后,在细胞内大部分被转化为还原性更强的二氢硫辛酸(DHLA)(Haramaki 等,1997)发挥作用。其抗氧化作用主要表现在以下的方面:①清除氧的自由基;②螯合氧化性的金属离子(Mn^{2+}、Cd^{2+}、Zn^{2+});③修复抗氧化损伤;④影响基因表达,保护 DNA 免受氧化损伤(Packer 等,1995);⑤还原产物二氢硫辛酸还具有保护抗氧化剂免受氧化和再生其他抗氧化剂(如维生素 C、维生素 E、谷胱甘肽)的作用。目前关于硫辛酸抗氧化的作用主要集中在陆生的脊椎动物(Arivazhagan 等,2002a,b;Chae 等,2008;Fujita 和 Hirofumi,2008)。Cd 优先和细胞中的巯基基团结合从而影响细胞的硫醇结构(Muller 和 Menzel,1988),含有巯基基团的物质被认为是最有效的抵抗重金属毒性的物质(Hatch 等,1978)。Sumathi(1996)通过小鼠体外实验研究谷胱甘肽、半胱氨酸、二硫苏糖醇、硫辛酸对 Cd 导致的肝脏自由基的清除,发现硫辛酸的效果最好。在前期的 Cd 攻毒实验中发现,GSH 在抵抗 Cd 的氧化应激中发挥了重要的作用。

第七章　饲料硫辛酸解除镉对皱纹盘鲍毒性作用的研究

但是目前在水产动物中关于硫辛酸抵抗重金属的氧化应激的实验还未见相关报道。在重金属污染越来越严重的今天，研究通过营养途径来降低重金属的毒性成为近年来的研究热点。本实验通过研究饲料中的硫辛酸对皱纹盘鲍抗 Cd 的作用及机制，为重金属解毒提供可靠的数据支持。

第二节　摄食生长实验设计

（一）实验动物和养殖管理设计

皱纹盘鲍购自青岛鳌山卫育苗场，为当年人工孵化的同一批鲍苗。在中国海洋大学水产馆养殖系统中暂养 2 周后挑选大小一致的健康个体［平均体重为（3.17±0.01）g］随机分成 3 组，每组 3 个重复，每个重复 50 只鲍。生长实验在中国海洋大学水产馆养殖系统中进行，静水养殖 2 个月。实验期间，每天换水两次，每次换水量为实验缸水量的一半。每天 17：00 投喂人工饲料，次日 8：00 清底，并密切观察采食及健康状况。养殖过程中水温 18~21℃，盐度为 22~28，pH 为 7.4~7.9，溶氧量大于 6mg/L。

（二）生长实验设计

根据前面的实验得知，重金属 Cd 对皱纹盘鲍的半致死浓度为 2.848mg/L。笔者团队选择半致死浓度的 1/8 作为本实验的浓度，即选择 0.35mg/L 作为本实验 Cd 的浓度。Cd 的实测值为（0.34±0.01）mg/L，Cd 的添加形式为 $CdCl_2 \cdot 2.5H_2O$。饲料中添加硫辛酸的水平为 0、700、2 100mg/kg，硒源为 $Na_2SeO_3 \cdot 5H_2O$。基础饲料配方参照 Mai 等（2003），为半精制饲料，蛋白质源为明胶和酪蛋白，脂肪源为鲱鱼油和大豆油，两者的比例为 1∶1，主要糖源为糊精，再辅以纤维素、维生素和矿物质（不含硒）等配制而成。基础饲料中的常规指标，包括粗蛋白质、粗脂肪和粗灰分的测定参照 AOAC（1995）。皱纹盘鲍的饲料配方及其营养成分见表 7-1。皱纹盘鲍饲料的具体配制步骤和保存方法参考 Zhang 等（2007）。

表 7-1 基础饲料配方及其营养成分

成分	含量（%）
酪蛋白[a]	25.00
明胶[b]	6.00
糊精[b]	33.50
羧甲基纤维素[b]	5.00
海藻酸钠[b]	20.00
维生素混合物[c]	2.00
矿物质混合物[d]	4.50
氯化胆碱[b]	0.50
豆油：鲱鱼油[e]	3.50
概略养分分析（以干重计）	
粗蛋白质	30.81
粗脂肪	3.85
粗灰分	11.01

注：a. sigma 公司。

b. 国药集团上海化学试剂有限公司。

c. 每1 000g饲料中含有：盐酸硫胺素，120mg；核黄素，100mg；叶酸，30mg；盐酸吡哆素，40mg；烟酸，800mg；泛酸钙，200mg；肌醇，4 000mg；生物素，12mg；维生素 B_{12}，0.18mg；维生素 C，4 000mg；维生素 E，450mg；维生素 K_3，80mg；维生素 A，100 000IU；维生素 D，2 000IU。

d. 每1 000g饲料中含有：NaCl，0.4g；$MgSO_4 \cdot 7H_2O$，6.0g；$NaH_2PO_4 \cdot 2H_2O$，10.0g；KH_2PO_4，20.0g；$Ca(H_2PO_4)_2 \cdot H_2O$，8.0g；Fe-柠檬酸，1.0g；$ZnSO_4 \cdot 7H_2O$，141.2mg；$MnSO_4 \cdot H_2O$，64.8mg；$CuSO_4 \cdot 5H_2O$，12.4mg；$CoCl_2 \cdot 6H_2O$，0.4mg；KIO_3，1.2mg。$Na_2SeO_3 \cdot 5H_2O$，1.0mg。

e. 豆油：鲱鱼油=1:1。

（三）取样与样品处理

养殖实验结束时，皱纹盘鲍禁食3d，排空其肠道内容物。实验缸中所有的皱纹盘鲍称重计数后，收集肝胰脏、肌肉、鳃、外套膜、贝壳。肝胰脏、肌肉、鳃和外套膜剪成小块混匀后分装在小管中，保存在-80℃冰箱中待测。离心得到的血清立即放在液氮中速冻后放入-80℃冰箱中保存。肝胰脏立即放入无 RNA 酶管中，液氮速冻后放入-80℃冰箱。

肝胰脏样品使用前解冻，加入预冷的0.86%生理盐水，冰上匀浆，然后4℃，4 000r/min离心15min，取上清测定抗氧化指标。

以特定增长率（SGR）衡量鲍的生长情况：

$$SGR = 100\ (\ln W_t - \ln W_i)\ /t$$

式中，W_t、W_i 分别代表鲍的终末体重和初始体重（g）；t 代表时间（d）。

第三节 饲料中的硫辛酸对镉胁迫下皱纹盘鲍生长存活和组织镉含量的影响

（一）对生长存活的影响

如表 7-2 所示，饲料中不同的 LA 添加水平对 Cd 胁迫下皱纹盘鲍生长和存活均无显著性影响（$P > 0.05$）。各组皱纹盘鲍的存活率在 90.38%～93.33%；各组生长没有显著性差异。

表 7-2 饲料中的硫辛酸（LA）对 Cd 胁迫下的皱纹盘鲍生长和存活的影响（平均值±标准误，$n = 3$）

饲料中 LA 含量 (mg/kg)	初始体重（g）	终末体重（g）	SGR（%）	存活率（%）
对照	3.14±0.01	3.81±0.12	0.32±0.32	0.93±0.02
700	3.16±0.01	3.75±0.04	0.28±0.28	0.92±0.04
2100	3.17±0.01	3.84±0.07	0.32±0.32	0.94±0.02
One-way ANOVA				
P 值	0.079	0.716	0.674	0.815
F 值	3.998	0.353	0.421	0.212

注：同列数据栏中，经 Turkey 检验差异不显著的平均值之间用相同的字母表示（$P > 0.05$）。

如表 7-3 所示，饲料中的 LA 显著性影响皱纹盘鲍血清、肌肉、外套膜、鳃和肝胰脏中的 Cd 含量（$P < 0.05$）。700、2 100 mg/kg 实验组血清、外套膜、鳃和肝胰脏中 Cd 含量显著低于对照组（$P < 0.05$）。2 100 mg/kg 实验组肌肉中的 Cd 含量显著低于对照组（$P < 0.05$），700 mg/kg 实验组肌肉中的 Cd 含量虽然低于对照组，但是无显著性差异（$P > 0.05$）。贝壳中 Cd 的含量在各实验组无显著性差异（$P > 0.05$）。

表 7-3 饲料中的硫辛酸 (LA) 对 Cd 胁迫下的皱纹盘鲍组织 Cd 含量的影响 (平均值±标准误, $n=3$)

饲料中 LA 含量 (mg/kg)	血清 (μg/mL)	贝壳 (μg/g)	肌肉 (μg/g)	外套膜 (μg/g)	鳃 (μg/g)	肝胰脏 (μg/g)
对照	2.23±0.04[a]	18.55±2.08	15.37±0.23[a]	33.48±1.20[a]	77.42±0.38[a]	426.11±3.63[a]
700	1.45±0.10[b]	20.25±4.58	12.71±1.01[ab]	21.76±2.54[b]	64.98±2.83[b]	209.93±21.08[c]
2100	1.38±0.15[b]	25.89±8.13	11.72±0.27[b]	24.81±0.32[b]	67.18±2.07[b]	321.08±11.00[b]
One-way ANOVA						
P 值	0.002	0.638	0.014	0.006	0.011	0.000
F 值	19.943	0.484	9.424	13.887	10.661	60.606

注: 同列数据栏中, 经 Turkey 检验差异不显著的平均值之间用相同的字母表示 ($P>0.05$)。

(二) 对组织金属含量的影响

本实验研究发现，700、2 100mg/kg LA 实验组降低了皱纹盘鲍血清、肝胰脏、肌肉、外套膜和鳃中的 Cd 含量。首先，LA 和其代谢产物 DHLA 具有螯合金属离子的能力（Ou 等，1995）。Biewenga 等（1997）报道，DHLA 通过螯合 Cd 来发挥其抗氧化作用，从而降低金属离子的积累。金属硫蛋白是动物体内重要的解毒蛋白，MT 与 Cd 等重金属结合可以减少其对组织的损害。LA 和 DHLA 的氧化还原反应中可以再生抗氧化剂 GSH，在本实验中，添加 LA 实验组的 GSH 显著高于对照组，而且有研究报道 GSH 的半胱氨基酸部分的巯基对 Cd 具有较高的亲和力，形成具有较高稳定能力的硫醇盐复合物，便于机体排出体外（Wang 和 Ballatori，1998）。猜测 MT 和 GSH 在 LA 降低机体 Cd 含量中起重要的作用，但是具体的机制还不清楚。

第四节 饲料中的硫辛酸对镉胁迫下皱纹盘鲍肝胰脏抗氧化和氧化指标的影响

(一) 对抗氧化指标的影响

如表 7-4 所示，各 LA 处理组对皱纹盘鲍肝胰脏 CAT、Se-GPx、GST 的活力和 GSH 的含量产生显著性差异（$P<0.05$）。700、2 100mg/kg 实验组的肝胰脏 GST、Se-GPx 和 GSH 的含量显著高于对照组（$P<0.05$）。700mg/kg LA 实验组的肝胰脏 CAT 活力显著高于对照组（$P<0.05$），2 100mg/kg LA 实验组的肝胰脏 CAT 高于对照组，但与对照组并无显著性差异（$P>0.05$）。700、2 100mg/kg 实验组的 SOD 含量均高于对照组，但和对照组无显著性差异（$P>0.05$）。各 LA 添加组对皱纹盘鲍肝胰脏的 TrxR、TrxP 的活性及 Trx 的含量无显著性差异（$P>0.05$）。

表 7-4 饲料中的硫辛酸(LA)对 Cd 胁迫下的皱纹盘鲍肝胰脏抗氧化相关指标的影响(平均值±标准误，$n=3$)

饲料中 LA 含量 (mg/kg)	SOD	CAT	Se-GPx	GST	GSH	Trx	TrxR	TrxP
0	11.94±0.16	3.01±0.30^b	8.47±0.30^b	1.99±0.20^b	8.80±0.12^a	298.47±1.39	2.71±0.11	6.52±0.64
700	12.54±0.38	3.96±0.21^a	9.85±0.18^a	3.57±0.11^a	9.69±0.22^b	294.76±0.75	2.64±0.24	6.81±0.35
2100	12.15±0.81	3.52±0.24^{ab}	10.03±0.03^a	3.81±0.03^a	10.44±0.09^c	294.38±1.90	2.85±0.16	5.66±0.43
One-way ANOVA								
P 值	0.724	0.01	0.003	0.000	0.001	0.161	0.7	0.298
F 值	0.341	10.707	18.237	55.115	24.042	2.515	0.379	1.494

注：SOD，超氧化物歧化酶(U，每毫克蛋白中)；CAT，过氧化氢酶(U，每毫克蛋白中)；GPx，谷胱甘肽过氧化物酶(U，每毫克蛋白中)；GST，谷胱甘肽硫转移酶(U，每毫克蛋白中)；GSH，还原性谷胱甘肽(mg，每克蛋白中)；Trx，硫氧还蛋白(U/mg Prot)；TrxR，硫氧还蛋白还原酶(mU/L)；TrxP，硫氧还蛋白过氧化物酶(U/mg Prot)。同列数据栏中，经 Turkey 检验差异不显著的平均值之间用相同的字母表示($P>0.05$)。

第七章 饲料硫辛酸解除镉对皱纹盘鲍毒性作用的研究

Cd 具有很强的致癌性，在我国的锦州湾、渤海湾、胶州湾其浓度超过了国家渔业水质标准（Xu 等，2000；Xu 等，2005；Chen 等，2004）。重金属 Cd 被证明可以导致 ROS 的产生（Di Giulio 等，1995；Tamás 等，2009），从而使水产贝类产生氧化应激（Company 等，2004）。α-硫辛酸是一种很强的抗氧化剂，与还原型代谢产物二氢硫辛酸一起在生物体内可以清除几乎所有的自由基和活性氧。本次实验结果表明，在 Cd 胁迫下，添加 LA 实验组的肝胰脏 CAT、Se-GPx、GST 的活性及 GSH 的含量显著升高，这可能与 LA 和 DHLA 的抗氧化能力有关。另外，在 LA 和 DHLA 的氧化还原过程中，可以还原生物体内其他的抗氧化剂，如 GSH、维生素 C、维生素 E 等，它们形成独特的抗氧化剂的再生循环网络，抵抗机体产生的过量自由基（Packer 等，2001）。谷胱甘肽抗氧化系统在 LA 抗 Cd 的过程中发挥了重要的作用。GSH 是体内重要的含巯基的抗氧化剂，它可以立即与 Cd 离子结合形成巯基复合物，是阻止 Cd 对组织损伤的第一道防线。DHLA 可以将胱氨酸还原为半胱氨酸，半胱氨酸被吸收速率比胱氨酸快 10 倍，这样有利于 GSH 的生物合成，因而 LA 干预能显著增加细胞内 GSH 水平（Hultherg 等，2002）。半胱氨酸已被证明在保护脑微粒体和解除 Cd 对肾细胞的毒性中具有重要的作用（Ochi 等，1988；Chetty 等，1992）。Se-GPx、GST 都是以 GSH 为底物，清除机体的过氧化氢和有机过氧化物，一般情况下 Se-GPx 的作用比 GST 大。但也有研究表明，当 Se-GPx 活性下降时，GST 可以起补偿作用（Bell 等，1986）。刘学忠等（2004）等研究表明，LA 可以穿过血脑屏障，提高组织中 Se-GPx 活性。Se-GPx 重要功能就是能够清除体内各种有机过氧化物和 H_2O_2，保护细胞免受过量自由基的毒害。实验表明，在 Cd 离子的胁迫下，LA 的添加可以提高机体的抗氧化水平，尤其是谷胱甘肽抗氧化系统在 LA 抗 Cd 的氧化胁迫中起重要的作用。

（二）对氧化指标的影响

在 Cd 胁迫过程中产生过量的活性氧可以对生物体的多种大分

子造成损害，可以使脂类中的不饱和脂肪酸氧化，从而产生脂质过氧化物，脂质过氧化物进一步分解成丙二醛，它们甚至使蛋白质氧化或 DNA 发生断裂等（刘瑞明等，1990）。镉可以引起 DNA 单链断裂，形成 DNA 碱基修饰产物（8-羟基脱氧鸟苷），并且损害 DNA 修复系统（Hartwig 等，1994）。本实验研究发现，添加 LA 的实验组降低 Cd 胁迫下皱纹盘鲍肝胰脏脂质过氧化水平及蛋白羰基化的水平、降低了 DNA 的断裂程度。这与 LA 可以显著提高 Cd 胁迫下的抗氧化力有关，而且 LA 和 DHLA 可以螯合汞、镉等金属离子，抑制其对生物体的产生的氧化损伤（Gerreke 等，1997；Lodge 等，1998）。

第五节 饲料中的硫辛酸对镉胁迫下肝胰脏金属硫蛋白（MT）含量、MT mRNA 和 MTF-1 mRNA 表达的影响

如表 7-5 所示，2 100mg/kg LA 实验组的肝胰脏 MDA 的含量显著低于对照组（$P<0.05$），700mg/kg LA 实验组的肝胰脏 MDA 的含量低于对照组，但是和对照组无显著性差异（$P>0.05$）。700、2 100mg/kg LA 实验组的肝胰脏蛋白羰基的含量显著低于对照组（$P<0.05$）。700、2 100mg/kg LA 实验组肝胰脏 DNA 断裂程度显著低于对照组（$P<0.05$）。

表 7-5 饲料中的硫辛酸（LA）对 Cd 胁迫下皱纹盘鲍肝胰脏抗氧化指标的影响（平均值±标准误，$n=3$）

饮料中 LA 含量	MDA (nmol，每毫克蛋白中)	蛋白质羰基化 (nmol，每毫克蛋白中)	DNA 断裂程度 （F 值）
对照	23.70±0.29[a]	9.40±1.11[a]	0.12±0.01[b]
700	21.12±0.61[a]	4.77±0.06[b]	0.30±0.03[a]
2100	16.04±0.86[b]	3.94±0.05[b]	0.37±0.05[a]

第七章 饲料硫辛酸解除镉对皱纹盘鲍毒性作用的研究

（续）

饮料中 LA 含量	MDA (nmol，每毫克蛋白中)	蛋白质羰基化 (nmol，每毫克蛋白中)	DNA 断裂程度 （F 值）
One-way ANOVA			
P 值	0.000	0.002	0.005
F 值	37.83	20.767	15.105

注：同列数据栏中，经 Turkey 检验差异不显著的平均值之间用相同的字母表示（$P>0.05$）。

如图 7-1 所示，700、2 100mg/kg 实验组的 MT 含量显著高于对照组（$P<0.05$），且 700mg/kg 实验组的 MT 含量显著高于 2 100 mg/kg（$P<0.05$）。700、2 100mg/kg 实验组的 MT mRNA 及 MTF-1 的表达显著高于对照组（$P<0.05$）。

MT 在抗重金属的应激中发挥重要的作用。首先，它可以通过巯基和自由基的结合，降低重金属过程中产生的过量 ROS。其次，MT 可以与多种的重金属结合，减轻重金属对生物体的氧化损伤。再次，MT 中的巯基基团可以通过氢供体使受损的 DNA 得以修复。本实验研究发现，饲料添加 700、2 100mg/kg LA 可以升高皱纹盘鲍肝胰脏 MT 的含量及 MT mRNA 的表达。最后，本实验还发现 MT 的调控因子 MTF-1 的 mRNA 的水平随着 MT mRNA 的升高而显著升高，可以推测金属硫蛋白的表达受 MTF-1 的调控，说明 MT 的这一调控路径在 LA 抗 Cd 的氧化胁迫及降低重金属的含量中发挥重要的作用，但是具体的机制还需要进一步的研究。

小结：

（1）饲料中添加 700、2 100mg/kg 的硒可以显著降低皱纹盘鲍肝胰脏、血清、鳃、外套膜和肌肉中的 Cd 组织含量，并可以降低皱纹盘鲍肝胰脏中脂质过氧化水平和蛋白羰基化水平及 DNA 的损伤程度，说明饲料中添加硫辛酸可以减轻 Cd 对皱纹盘鲍的氧化损伤。

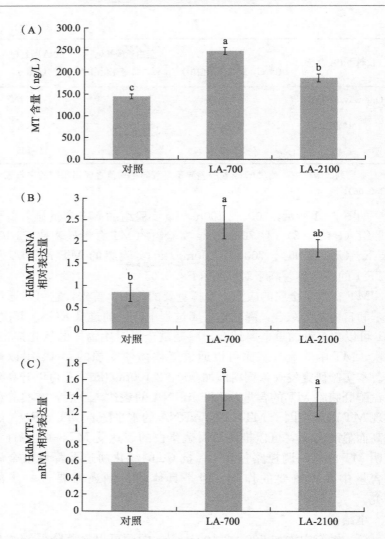

图7-1 不同硫辛酸（LA）处理对皱纹盘鲍Cd胁迫下的肝胰腺中MT含量、HdhMT和HdhMTF-1基因的表达的影响（平均值±标准误，$n=3$）

(A) 不同浓度下的LA对Cd胁迫下皱纹盘鲍肝胰脏MT含量的影响；(B) 不同浓度下的LA对Cd胁迫下皱纹盘鲍肝胰脏MT mRNA表达的影响；(C) 不同浓度下的LA对Cd胁迫下皱纹盘鲍肝胰脏MTF-1 mRNA表达的影响。

(2) 饲料中添加 700、2 100mg/kg 的硫辛酸显著升高了皱纹盘鲍肝胰脏抗氧化酶 CAT、Se-GPx、GST 的活性及 GSH 的含量，说明硫辛酸在一定程度可以提高 Cd 胁迫下的皱纹盘鲍肝胰脏的抗氧化水平。

(3) 饲料中硫辛酸的添加显著升高了 Cd 胁迫下的皱纹盘鲍肝胰脏金属硫蛋白的含量及表达，而且显著升高了 MTF-1 mRNA 的表达。表明在硫辛酸缓解 Cd 毒性的过程中，MT 发挥重要作用，具体机制还需要继续研究。

参 考 文 献

李奕，2006. α-硫辛酸对耐力训练小鼠力竭运动后谷胱甘肽抗氧化系统的影响 [D]. 上海：华东师范大学.

刘学忠，崔旭，卞建春，等，2004. 硫辛酸在大鼠全脑缺血再灌注损伤中的神经保护作用 [J]. 中国兽医学报，24（4）：388-390.

田芳，仲伟鉴，应贤平，等，2007. α-硫辛酸对 H_2O_2 诱导的细胞活性氧水平及 DNA 氧化损伤的影响 [J]. 环境与职业医学，24（2）：180-189.

ARIVAZHAGAN P, SHILA S, KUMARAN S, et al, 2002a. Effect of DL-α-lipoic acid on the status of lipid peroxidation and protein oxidation in various brain regions of aged rats [J]. Exp. Geront, 37（6）：803-811.

ARIVAZHAGAN P, SHILA S, NARCHONAI E, et al, 2002b. α-Lipoic Acid Enhances Reduced Glutathione, Ascorbic Acid, and α-Tocopherol in Aged Rats [J]. J. Anti-Aging Med, 5（3）：265-269.

BELL JG, ADRON JW, COWEY CB, 1986. Effect of selenium deficiency on hydroperoxide-stimulated release of glutathione from isolated perfused liver of rainbow trout (*Salmo gairdneri*) [J]. Br. J. Nutr, 56：421-428.

BONOMI F, CERIOLI A, PAGANI S, 1989. Molecular aspects of the removal of the removal of ferritin-bound iron by DL-dihydrolipoate [J]. Biochim. Biophys. Acta, 994（2）：180-186.

BUSSE E, ZIMMER G, SCHOPOHL B, 1992. Influence of α-lipoic acid on intracellular glutathione in vitro and in vivo [J]. Arzneimttel-Forschung, 42（6）：829-831.

CAO Z, TSANG M, ZHAO H, et al, 2003. Induction of endogenous antioxidants and Phase 2 enzymes by a-lipoic acid in rat cardiac H9C2 cells: protection against oxidative injure [J]. Biochem. Biophys. Res. Commun, 310: 979-985.

CAREY A, WILLIAMS, RHONDA M, et al, 2002. Lipoic acid as an antioxidant in mature thoroughbred geldings: A Preliminary Study [J]. J. Nutr, 2: 1628S-1631S.

CHAE C H, SHIN C H, KIM H T, 2008. The combination of α-lipoic acid supplementation and aerobic exercise inhibits lipid peroxidation in rat skeletal muscles [J]. Nutrition Research, 28 (6): 399-405.

CHELLU S, CHETTY, et al, 1992. The effects of cadmium in vitro on adenosine triphosphatase system and protection by thiol reagents in rat brain microsomes [J]. Arch. Environ. Con. Tox, 22 (4): 456-458.

COMPANY R, SERAFIM A, BEBIANNO MJ, et al, 2004. Effect of cadmium, copper and mercury on antioxidant enzyme activities and lipid peroxidation in the gills of the hydrothermal vent mussel *Bathymodiolus azoricus* [J]. Mar. Environ. Res, 58: 377-381.

DEVASAGAYAM TPA, SUBRAMANIAN M, PRADHAN DS, et al, 1993. Prevention of singlet oxygen-induced DNA damage by lipoate [J]. Chem. Biol. Interact, 79-92.

EASON RC, ARCHER HE, AKHTAR S, et al, 2002. Lipoic acid increases glucose uptake by skeletal muscles of obese-diabetic ob/ob mice [J]. Diabetes Obes. Metab, 4 (1) : 29-35.

FUJITA, HIROFUMI, 2008. α-Lipoic acid suppresses 6-hydroxydopamine-induced ROS generation and apoptosis through the stimulation of glutathione synthesis but not by the expression of heme oxygenase-1 [J]. Brain Research, 1206 (1): 1-12.

GABBIANELLI R, LUPIDI G, VILLARINI M, et al, 2003. DNA damage induced by copper on erythrocytes of gilthead sea bream*Sparusaurata* and mollusk *Scapharcainaequivalvis* [J]. Arch. Environ. Contam. Toxicol, 45: 350-356.

GERREKE PH, BIEWENGA, GUIDO R, 1997. The pharmacology of the antioxidant lipoic acid [J]. Gen. Pharmac, 29 (3): 315-331.

HALLIWELL B, GUTTERIDGE MC, 1984. Oxygen toxicity, oxygen radicals, transition metals and disease [J]. Biochem. J, 219: 1-14.

HATCH R T, MENAWAT A, 1978. Biological removal and recovery of trace heavy metals [J]. Biotechnol. Bioeng. Symp. (United States), 8: 191-203.

HARAMAKI N, HAN D, HANDELMAN G J, et al, 1997. Cytosolic and mitochondrial systems for NADH-and NADPH-dependent reduction of alpha-lipoic acid. [J]. Free Radical Bio. Med, 22 (3): 535-542.

HARTWIG A, KRÜGER I, BEYERSMANN D, 1994. Mechanisms in nickel genotoxicity: the significance of interactions with DNA repair [J]. Toxicology Letters, 72 (1-3): 353.

HARRIS ED, 1991. Copper transport: an overview [J]. Exp. Biol. Med, 196 (2): 130-140.

JE JH, LEE TH, KIM DH, et al, 2008. Mitochondrial ATP synthase is a target for TNBS-induced protein carbonylation in XS-106 dendritic cells [J]. Proteomics, 8: 2384-2393.

JIMÉNEZ I, ARACENA P, LETELIER ME, 2002. Chronic exposure of HepG2 cells to excess copper results in depletion of glutathione and induction of metallothionein [J]. Toxicology in vitro, 16 (2): 167-175.

JIMÉNEZ I, SPEISKY H, 2000. Effects of copper ions on the free radicalscavenging properties of reduced glutathione: implications of a complex formation [J]. Trace Elem. Biol. Med, 14: 161-167.

KAPPUS H, 1985. Lipid peroxidation: mechanisms, analysis, enzymology and biological relevance [M]. //SIES H. Oxidative Stress. London: Academic Press.

KOWLURU RA, ODENBACH S, BASAK S, 2005. Long-term administration of lipoic acid inhibits retinopathy in diabetic rats via regulating mitochondrial superoxide dismutase [J]. Invest Ophthalmol. Vis. Sci, 46: 396-422.

LEONARD P, RYBAK, KAZIM HUSAIN, 1999. Dose dependent protection by lipoic acid against cisplatin-induced ototoxicity in rats: antioxidant defense system [J]. Toxicol. SCI, 47 (3): 195-202.

LODGE JK, TRABER MG, PACKER L, 1998. Thiol chelation of Cu^{2+} by dihydrolipoic acid prevents human low density lipoprotein peroxidation [J]. Free Rad. Med, 25 (3): 287-297.

MAI K, ZHANG W, TAN B, et al, 2003. Effects of dietary zinc on the shell biomineralization in abalone *Haliotis discus hannai* Ino [J]. J. Exp. Mar. Biol. Ecol, 283: 51-62.

MONSERRAT JM, LIMA JV, FERREIRA JLR, et al, 2008. Modulation of antioxidant and detoxification responses mediated by lipoic acid in the fish *Corydoras paleatus* (Callychthyidae) [J]. Comp. Biochem. Physiol, Part C, 148: 287-292.

MULLER L, MENZEL H, 1988. Studies on the efficacy of lipoate and dihydrolipoate in the alteration of cadmium^{2+} toxicity in isolated hepatocytes [J]. BBA. Mol. cell. 1052 (3): 386-391.

NAVARI IZZO F, QUARTACCI MF, SGHERRIC C. LIPOIC, 2002. Acid: a unique antioxidant in the detoxification of activated oxygen species. Plant Physiol. Biochem, 40: 463-470.

OCHI T, OTSUKA F, TAKAHASHI K, et al, 1988. Glutathione and metallothioneins as cellular defense against cadmium toxicity in cultured Chinese hamster cells [J]. Chem. Biol. Interact, 65: 1-14.

PACKER L, KRAEMER K, RIMBACH G, 2001. Molecular aspects of lipoic acid in the prevention of diabetes complications [J]. Nutrition, 17 (10): 888-895.

PACKER L, WITT EH, TRITSCHLER HJ, 1995. Alpha-lipoic acid as a biological antioxidant [J]. Free Radic. Biol. Med, 19 (2): 227-250.

PALACEA VP, BROWN SB, BARON CI, et al, 1998. An evaluation of the relationships among oxidative stress, antioxidant vitam ins and early mortality syndrome (EMS) of lake trout (*Salvelinus namaycush*) from Lake Ontario [J]. Aquat. Toxicol, 43: 259-268.

QUINN JF, BUSSIERE JR, HANLLNOND RS, et al, 2007. Chronic dietary α-lipoic acid reduces deficits in hippocampal memory of aged Tg 2576 mice [J]. Neurobiol. Aging, 28: 213-225.

REED, LESTER J, 2001. A trail of research from lipoic acid to alpha-keto acid dehydrogenase complexes [J]. J. Biol. Chem, 276 (42): 38329-38336.

RICHARDS M, RECENT, 1989. Recent developments in trace element metabolism and function: role of metallthionein in copper and zinc metabolism. J. Nutr, 119: 1062-1070.

RUAS CBG, CARVALHO CS, DE Araujo HSS, 2008. Oxidative stress biomarkers of exposure in the blood of cichlid species from a metal-contaminated river [J]. Ecotoxicol. Environ. Saf, 71: 86-93.

SHACTER E, WILLIAMS JA, LIM M, et al, 1994. Differential susceptibility of plasma proteins to oxidative modification: examination by western blot immunoassay [J]. Free Radical Biol. Med, 17: 429-437.

SUMATHI R, 1996. Effect of DL α-lipoic acid on tissue redox state in acute cadmium-challenged tissues [J]. Journal of Nutritional Biochemistry, 7 (2): 85-92.

TAMÁS L, VALENTOVICOVÁ K, HALUSKOVÁ L, et al, 2009. Effect of cadmium on the distribution of hydroxyl radical, superoxide and hydrogen peroxide in barley root tip [J]. Protoplasma, 236: 67-72.

THORNALLEY PJ, VASAK M, 1985. Possible role for metallothionein in protection against radiation-induced oxidative stress. Kinetics and mechanism of its reaction with superoxide and hydroxyl radicals [J]. Biochim. Biophys. Acta, 827: 36-44.

VIARENGO A, CANESI L, PERTICA M, et al, 1990. Heavy metal effect on lipid peroxidation in the tissues of *Mytilus galloprovincialis* L [J]. Comp. Biochem. Physiol (C), 97: 37-42.

WANG HW, XU HM, XIAO GH, 2010. Effects of selenium on the antioxidant enzymes response of Neocaridinaheteropodaexposed to ambient nitrite. Bull [J]. Environ. Contam. Toxicol, 84: 112-117.

WANG W, BALLATORI N, 1998. Endogenous glutathione conjugates: occurrence and biological functions. Pharmacol Rev 50: 335-356 [J]. Pharmacol. Rev, 50 (3): 335-356.

ZHANG L, XING GQ, BARKER JL, 2001. α-lipoic acid protects rat cortical neurons against cell death induced by amyloid and hydrogen peroxide through the Akt signaling pathway [J]. Neurosci. Lett, 312 (3): 125-128.

ZHANG W, MAI K, XU W, et al, 2007. Interaction between vitamins A and D on growth and metabolic responses of abalone *Haliotis discus hannai*, Ino [J]. J. Shellfish Res, 26: 51-58.

图书在版编目（CIP）数据

抗氧化剂对皱纹盘鲍重金属的解毒作用研究 / 类延菊著. —北京：中国农业出版社，2022.9
ISBN 978-7-109-29901-6

Ⅰ.①抗… Ⅱ.①类… Ⅲ.①抗氧化剂－作用－海洋生物－贝类养殖－重金属中毒－研究 Ⅳ.①S968.31

中国版本图书馆 CIP 数据核字（2022）第 157386 号

中国农业出版社出版
地址：北京市朝阳区麦子店街 18 号楼
邮编：100125
责任编辑：周锦玉　文字编辑：陈睿赜　苏倩倩
版式设计：杜　然　责任校对：吴丽婷
印刷：北京科印技术咨询服务有限公司
版次：2022 年 9 月第 1 版
印次：2022 年 9 月北京第 1 次印刷
发行：新华书店北京发行所
开本：880mm×1230mm　1/32
印张：4.5
字数：150 千字
定价：25.00 元

版权所有·侵权必究
凡购买本社图书，如有印装质量问题，我社负责调换。
服务电话：010 - 59195115　010 - 59194918